T0211773

Metaphysics from a Biological Point of View

Also by Stephen Boulter

THE REDISCOVERY OF COMMON SENSE PHILOSOPHY

Metaphysics from a Biological Point of View

Stephen Boulter
Oxford Brookes University, UK

First published 2013 by
PALGRAVE MACMILLAN

Palgrave Macmillan in the UK is an imprint of Macmillan Publishers Limited,
registered in England, company number 785998, of Houndmills, Basingstoke,
Hampshire RG21 6XS.

Palgrave Macmillan in the US is a division of St Martin's Press LLC,
175 Fifth Avenue, New York, NY 10010.

Palgrave Macmillan is the global academic imprint of the above companies
and has companies and representatives throughout the world.

Palgrave® and Macmillan® are registered trademarks in the United States,
the United Kingdom, Europe and other countries

ISBN 978-1-349-45861-5   ISBN 978-1-137-32282-1 (eBook)
DOI 10.1057/9781137322821

This book is printed on paper suitable for recycling and made from fully
managed and sustained forest sources. Logging, pulping and manufacturing
processes are expected to conform to the environmental regulations of the
country of origin.

A catalogue record for this book is available from the British Library.

A catalog record for this book is available from the Library of Congress.

Contents

Acknowledgements

I have benefited enormously from feedback received to papers delivered at the following conferences: 'The Cartesian "Myth of the Ego": The Analytic/Continental Divide', University of Nijmegen, 2010; 'Metaphysics: Aristotelian, Scholastic and Analytic', Charles University, Prague, 2010; 'Metaphysics and the Philosophy of Science', University of Toronto, 2011; 'What Is Really Possible?' University of Utrecht, 2011; 'Science and Metaphysics', University of Kent, 2012; and 'Analytical Approaches to Thomas Aquinas', Jagiellonian University, Krakow, 2012. I want to thank Edward Feser and David Oderberg for comments on early versions of Chapters 2 and 6 respectively.

I would also like to thank my wife, Eileen Reid, whose strength in difficult times has been inspirational.

Introduction

This book sets out and defends a collection of distinct but interlocking theses across a number of different debates. Although predominantly about metaphysical matters, the positions adopted here are largely motivated by a particular meta-philosophical view regarding the role of philosophy within the general intellectual economy. This meta-philosophy is crucial to what follows, for it forms the basis of the claim that metaphysics is indispensible to the business of philosophy. That is, if philosophers *qua* philosophers are to play their useful part in the wider intellectual economy, they will have to engage at some point in metaphysical reflection. This claim will strike many as implausible, for it has been a commonplace since Kant to question the very possibility, let alone the necessity, of metaphysics. But a further aspect of the overarching meta-philosophy is a key methodological claim that the role envisaged for philosophy within the intellectual economy provides the criterion against which competing metaphysical theses are to be measured. In short, it is possible, *pace* Kant and other doubters, to provide rational warrants for substantial metaphysical claims about the fundamental structure of reality as a whole. These matters regarding the necessity and possibility of metaphysics are the subject of Part I.

In Part II the approach to metaphysics presented in Part I is deployed in the defence of particular metaphysical theses. In general I defend a version of metaphysical realism. Essentially this is the view that reality contains, among other things, mind-independent entities whose natures and properties owe nothing to our thought or representations of them. More particularly, I defend a metaphysics of common sense.

That is, I defend the view that we must accept in all ontological seri-
ousness the mind-independent existence of the sorts of middle-size
objects Moore insisted upon in his celebrated paper 'A Defence of
Common Sense'.[1] More particularly still, and certainly more contro-
versially, I defend what I take to be the most sophisticated metaphysics of common sense developed to date – that of the Scholastic
Aristotelians. But as my title suggests, this effort in metaphysics does
not follow the usual pattern. For reasons developed in Part I, I take it
that any metaphysical thesis is warranted to the extent that it provides
a solution to problems emerging from the first-order disciplines. It is
for this reason that this foray into metaphysics focuses extensively on
ostensibly extra-philosophical matters – in this case puzzles arising
out of biology. What ultimately emerges from these efforts is a defence
of Scholastic metaphysical principles based on contemporary biological theory, an irony not likely to be lost on anyone.

Of all the claims advanced in this book perhaps the most quixotic is
the defence of Scholasticism. However, a growing number of metaphysicians working within the broadly analytic tradition are coming to
recognise the enduring value of the Aristotelian framework.[2] Moreover,
a fair few of this happy band have recognised that much that is left
vague and underdeveloped in Aristotle was expanded upon and refined
to an astonishing degree of sophistication by the medieval schoolmen.[3] Thus any contemporary metaphysician with an Aristotelian
bent will find much to admire in the Scholastics. Nonetheless, to ease
the contemporary reader into what is likely to be a rather foreign frame
of mind, I begin in Chapter 1 with some historical observations on
the twinned fate of Scholasticism and metaphysics. An understanding
of the reasons for the decline of Scholasticism and the not unrelated
decline of the prestige of metaphysics in the modern period goes some
way to explaining why one might recommend a return to premodern
modes of philosophising. I then turn in Chapter 2 to the main business of Part I – the necessity and possibility of metaphysics, and the
exposition of the aporetic method. The main claim of this method
is that one recognises the correctness of a metaphysical principle by
noting the facility with which it allows one to resolve an aporia. I close
Part I with a budget of aporia based on reflections on evolutionary
biology, and a brief statement of the key Scholastic principles to be
examined from a biological point of view.

Attention then turns in Part II to the aporia themselves. The first focuses on the provision of a theory of individuation for biological organisms, a theory that will at once support the claim that biological organisms must be given a place within one's ontology, and allow one to demarcate clearly between organisms, their parts and the colonies or groups they may join. The second aporia begins with the observation that biological individuals are subject to change. Metaphysicians have struggled to accommodate the reality of change without appealing to the distinction between essential and accidental properties. But essentialism is allegedly incompatible with evolutionary theory. Squaring this circle is the second challenge. The third aporia deals with the nature of explanation. Explanations in evolutionary biology trade on the ability to distinguish between the biologically necessary, contingent, possible and impossible. But there is currently no consensus on how to map the biological modalities. This lack of consensus undermines the explanatory credentials of evolutionary biology as they rely on distinctions that cannot as yet be drawn in a principled fashion. Addressing this lacuna is the third challenge. Finally, there is the challenge of incorporating the biological sciences into a wider view of the human condition embracing ethics and politics. Some have argued that biology can be the foundation of ethics. Others insist that the ethical is possible *despite* our biology. Respecting what seems right in both camps is the final challenge to be examined here. But the main claim of Part II is that the smoothest resolution of these problems is to be had by respecting Scholastic metaphysical principles.

This book incorporates material that has previously seen the light of day in various locations as standalone pieces. Chapter 1 employs material that appeared in 'The Medieval Origins of Conceivability Arguments', *Metaphilosophy*, Vol. 42, No. 5, 617–641, 2011; Chapter 2 incorporates material that appeared in an edited collection entitled *Aristotle on Method and Metaphysics* (Palgrave Macmillan, 2013); material in Chapter 5 first appeared in *Human Nature: Royal Institute of Philosophy Supplement* (70), 83–103, 2012, under the title 'Can Evolutionary Biology Do without Aristotelian Essentialism?'; Chapter 6 incorporates material that first appeared in a special collection of *Ratio on Classifying Reality*, Vol. XXV, No. 4, 425–441, 2012. This material appears with the kind permission of the editors.

Part I
On the Possibility of Metaphysics

1
A Return to Scholastic Metaphysics

Introduction

One of the most remarkable features of mid-to-late-twentieth-century analytic philosophy is the revival of interest in the traditional problems of metaphysics. It is true that philosophers never really abandoned metaphysics, at least in the sense that they never stopped making metaphysical claims of one sort or another. But it is undoubtedly the case that, at least officially, metaphysics was in very bad odour in mainstream circles prior to Kripke's (1972) rediscovery of necessary a posteriori propositions. For ever since Hume consigned metaphysics to the flames, and Kant asked how synthetic a priori propositions are possible, philosophers had grown accustomed to the idea that no profit is to be had from metaphysical reflection as traditionally understood. Metaphysics, once the queen of the sciences, was exposed by these Enlightenment thinkers as an impossibility, and its practitioners, however well intentioned, deluded. The passage of time seemed to confirm this assessment, and the only matter of dispute was how one should react to the death of metaphysics. Some, Kant for instance, emphasised the tragic aspect of metaphysics' calamities, for we cannot but ask metaphysical questions even when we know they cannot be answered (1965, p. 7). Others positively welcomed the demise of metaphysics as it was nothing more than a mask concealing a will to power that needed to be exposed for what it was.[1] Still others were more sanguine, maintaining that philosophy can get along just fine without metaphysics, either because philosophy has other business to attend to,[2] or because our legitimate metaphysical needs, if there are any, can be addressed by the sciences.[3]

But more recent history seems to corroborate Gilson's claim that 'metaphysics always buries its undertakers'.[4] Although we have yet to see anything like a restoration of metaphysics to its former glory, it is now a burgeoning research area, showing all the signs of returning to rude good health. However, given the long period of neglect from which metaphysics is only now emerging, it is perhaps not surprising that philosophers in the analytic tradition are still collectively finding their metaphysical feet, particularly when it comes to meta-metaphysical and methodological matters.[5] And, as we shall see in the next chapter, many still doubt there are any metaphysical feet to find. Thus contemporary metaphysicians find themselves in an exciting, but ultimately unstable position: Many of us are now convinced that metaphysical questions are worth pursuing, but we still lack a convincing story to offer our detractors about how we conduct our business *qua* metaphysicians. This book is an attempt to provide such a story.

And the core recommendation is this: Contemporary metaphysicians need not reinvent the discipline out of whole cloth. Much time and energy can be saved if we are willing to learn from the past masters, particularly the Scholastics. Aquinas, Scotus, Ockham and Suarez, the leading figures from the golden and silver ages of Scholasticism, provide plausible answers to our meta-metaphysical and methodological questions if only we are willing to listen. They also have tenable answers to first-order metaphysical questions, answers that deserve to be seen as serious contributions to our ongoing efforts at metaphysical reflection.

This recommendation is hard for many to swallow. While it is often a sensible tactic in chess to retreat the better to advance, this is not a standard move in philosophical contexts. Moreover, like Hobbes, most analytic philosophers labour under the false impression that 'the Schoolmen' were purveyors of a 'dark' and 'vain' philosophy expressed in 'insignificant Traines of strange or barbarous words' about obscure metaphysical entities justified only by empty ratiocinations and verbal disputations carried out under the suspicious eye of an ever vigilant Church.[6] It would take a book length study to explain how this gross caricature came to be so widely accepted amongst philosophers from the early modern period down to the present day.[7] But whatever the history, the view is false. The Scholastics were highly sophisticated metaphysicians, logicians, epistemologists, cognitive psychologists and ethicists, and there is much to be gained by a sustained study of their work. And far from being monolithic defenders of the faith, who 'resolve of their Conclusions before they know their Premises', the Scholastics knew very well how to distinguish their theology from their philosophy. That there is a clear distinction between Scholastic theology and Scholastic metaphysics is not always appreciated. This is no doubt due to the fact that their metaphysical discussions are usually intertwined with theological discussions, and because they rarely developed metaphysical views for their own sake. But this is an accidental and not an essential feature of their metaphysical reflections. It is perhaps Suarez's greatest contribution to the history of philosophy that he presented the fruits of Scholastic metaphysical reflection in a systematic fashion *independently* of any theological considerations in his monumental *Metaphysical Disputations*. That this was possible at all speaks to the fact that metaphysics had an independent life of its own even in the hands of theologians.

However, a couple of additional points are worth making in order to sweeten the suggestion that the Scholastics have much to teach us. First, although this book is a defence of metaphysics as traditionally understood and practised by the Scholastics, there is one important difference – the point of departure for metaphysical reflection. The great Scholastic metaphysicians were theologians first and foremost, and their metaphysical doctrines emerged in the course of their efforts to deal with theological problems. Theology plays no such role here. However, the sciences throw up problems that are structurally analogous to those tackled by the Scholastics. The account of metaphysics to be developed here is thus inspired by the meta-metaphysics and methodology of the Scholastics, but begins with problems emerging from the sciences rather than theology. The problems discussed in Part II are all drawn from reflection on the life sciences in particular, and so one can justly call this book an effort in metaphysics from a biological point of view. If anyone should baulk at this version of Scholasticism, it cannot be because it is compromised by theological commitments.

The second sweetener is the primary thesis of this introductory chapter, namely, that the fall of traditional metaphysics at the hands of Hume and Kant is not unrelated to developments in thirteenth- and fourteenth-century *theology*. Understanding the history of how Scholastic theology directly contributed to the fall of metaphysics as traditionally understood goes some way to explaining why Scholastic metaphysics is worth revisiting. Several key lessons emerge from this (necessarily brief) historical study: First, that Hume and Kant's critiques of metaphysics do not apply to metaphysics *per se*, but only to the metaphysical systems of their Early Modern contemporaries; and second, that the motivations that lead to the novelties of the Early Modern systems are theological, have little to do with philosophy *per se* and so need no longer act as constraints upon our own ongoing efforts at metaphysical reflection. Thus, a metaphysics shorn of these particular theological influences would be untouched by Humean and Kantian considerations. Finally, surprising as it sounds, such a metaphysics is more readily discerned amongst the early and middle Scholastics than amongst contemporary analytic metaphysicians, many of whom still labour under the influence of Carnap and Quine. The irony is that Carnap and Quine betray the influence of thirteenth- and fourteenth-century theology more so than an Aquinas, and so it is to the latter rather than the former that one ought to go for insights into the nature of metaphysics. Making good this claim is the primary business of this chapter.

Whatever happened to metaphysics?

The Ancient Greek philosophers happily engaged in metaphysical reflection, as did the Scholastics and the greats of the Early Modern period. But Hume and Kant convinced us that metaphysics, as traditionally understood, is not possible. This raises an obvious question: Why did philosophers of the calibre of Plato, Aristotle, Aquinas, Descartes, Spinoza and Leibniz – to name only a few luminaries – believe that metaphysics was worth pursuing, only for this view to fall so decisively from favour? What philosophical discoveries led to the negative assessment of the viability of metaphysics we find in Hume and Kant? These are not idle historical questions. Understanding the fall of metaphysics helps situate the question of the necessity and possibility of metaphysics in its proper context, allowing one to appreciate clearly what assumptions lie behind the Humean/Kantian consensus, assumptions that remain quietly at work in many quarters to this day.

As with all historical explanations of complex phenomena, the correct account is bound to be multifaceted, making appeal to a variety of explanatory factors. I want to focus on just one such factor, although I think it played a central part in our story. Unearthing this explanation requires an excursion into the unfamiliar and at times forbidding territory of the Middle Ages. But the excursion is worth the effort. For what emerges from such a study is that the targets of Hume and Kant's original critiques were primarily the rationalist, deductivist metaphysicians of the Early Modern period whose approaches to the discipline were significantly different from those adopted by Aristotle and the Scholastics. So our historical question can be made more precise by asking why metaphysics in the hands of Descartes, Spinoza and Leibniz is a rather different creature from that found amongst the Ancients and the Scholastics. Understanding the transition from the metaphysics of the Scholastics to the metaphysics of the Early Modern period proves crucial to an understanding of the fate that eventually befell the queen of the sciences.

The picture I am offering, in a nutshell, is as follows: With the Ancient Greeks, particularly Aristotle, and with the early and middle Scholastics, we find metaphysics in good health. The discipline has a purpose within philosophy as a whole, it has a generally agreed methodology and there is a general consensus that its questions are well formed and that warranted answers can be provided. However, this framework operated with certain modal assumptions, the most important of which is that reality displays forms of necessity of a non-logical variety. For theological reasons to be discussed later, this assumption was challenged in the late thirteenth century. Subsequently, metaphysicians, both Late Scholastic and Early Modern, continued to prosecute the original metaphysical project while accepting a distinct doctrine regarding the nature of necessity. This led to the adoption of forms of rationalism that had never been seriously entertained in Scholastic circles, and the result was the systems we associate with Descartes, Spinoza and Leibniz. Perhaps not surprisingly, these systems eventually came under severe, and largely warranted, attack. But the only alternatives to rationalism visible at the time were either to give up on metaphysics altogether, as Hume suggests, or to radically reconceive the discipline, as does Kant.

But all parties to this dispute adopted the modal assumption foreign to the original framework employed by traditional metaphysicians. And, up until Kripke's *Naming and Necessity*, most philosophers in the orthodox cannon have followed suit. The suggestion then is that it remains to be seen if there was anything amiss with metaphysics prior to the adoption of the theologically supported modal assumptions. I will be arguing that there was in fact nothing amiss with such a metaphysics. And as the suspect modal assumptions are no longer taken for granted, the option of revisiting the original project is open to us. I believe it is an option we should thoroughly explore.

I begin with a sketch of the general metaphysical project as traditionally conceived. In the first instance this means providing a clear picture of the formal objects of metaphysical reflection, i.e., those aspects of reality taken to fall within the domain of metaphysics as opposed to any of the other theoretical sciences. But just as important for present purposes is to be clear about the nature of the knowledge metaphysicians traditionally sought. As was the case in all the theoretical sciences, the goal towards which the metaphysician was working was the provision of certain and infallible knowledge of the formal objects of the discipline. But, according to Ancients and Moderns alike, the possibility of certainty depends not just on our cognitive ability adequately to track reality, but also on the modal nature of the object of knowledge. In particular, it was held that certain knowledge or *scientia* is possible only of those features of reality that 'cannot be otherwise'. These 'eternal and necessary truths', the *aeternae et necessariae veritates*, were the goal of metaphysical reflection.

In the next section I go on to show how *theologically* imposed changes to attitudes regarding the modal nature of reality brought about a radical restriction of the range of items with the modal features required to ground the possibility of certainty. This in turn had methodological implications that constrain how one can proceed with one's metaphysical reflections. As we shall see, if metaphysicians continue to accept that certainty is possible only of those things which 'cannot be otherwise', and they continue to accept the theologically motivated views on modality, then only three approaches to metaphysical reflection are open: One can adopt the rationalism of a Descartes, Spinoza and Leibniz; or, with Hume, one can assert that reality is radically contingent, and that this contingency undercuts the very possibility of metaphysics; or, with Kant, one can accept a

new and previously unrecognised source of necessity, viz., the struc-
turing mind, and shift one's attention from the noumenal (because
unstructured) to the phenomenal (because structured) world. The
value of this history is that it shows how one can avoid choosing
among these three unattractive approaches to metaphysics and their
contemporary manifestations.

The metaphysical project

Metaphysics as traditionally conceived, and as it will be understood
throughout this book, is the study of the most fundamental structure
of reality. In this conception a completed metaphysics would provide:
(i) an ontology, i.e., a complete catalogue of the most general and
irreducible categories of entities included in the domain of our most
unrestricted quantifiers;[8] (ii) an account of the relations that obtain
among entities of the various categories;[9] and (iii) an account of the
'theatre' in which many if not all entities have their being, namely,
space and time. It is important to note that metaphysicians traditionally do not suppose this study to be straightforwardly empirical in
nature if only because it cannot be assumed without further argument that all that exists exists in space and time – abstract objects
being one obvious possible candidate, immaterial minds another.
Consequently, the need to justify in a non-empirical fashion at least
some claims concerning the fundamental structure of reality is built
into the very conception of the traditional metaphysical project.[10]

　This is the overarching project of metaphysics as traditionally
conceived. Now while many of the pre-Socratics, and certainly
Plato, clearly considered such matters, it is perhaps fair to say that
the project as traditionally conceived clearly comes into view with
Aristotle's *Categories*. And it is worth pointing out that not all philosophers with metaphysical views engage in all aspects of the project.
Many prefer to focus on particular questions of interest to them,
such as the nature of causation, the existence of God or numbers,
mind/body relations, etc. We will have occasion to focus on particular topics in Part II, but in Part I when I refer to metaphysics it is this
overarching project I have in mind unless otherwise specified.

　But the account of metaphysics provided thus far is incomplete
in at least one important respect because it does not make explicit
a shared expectation regarding the formal objects of metaphysics,

an expectation shared by metaphysicians from Aristotle through to Kant.[11] Ever since Aristotle made it explicit in the *Posterior Analytics*, philosophers have prioritised knowledge, *not* of the *contingent* facts or aspects of reality, i.e., those aspects subject to change, but of those aspects of reality that cannot be otherwise.[12] This prioritisation of knowledge of what is necessary over knowledge of contingencies applies in all the theoretical sciences, and *a fortiori* in metaphysics. It is worth labouring this point because of its centrality to the history I will be tracing in this chapter. Aristotle writes in the *Metaphysics*:

> ...we must first say regarding the accidental, that there can be no scientific treatment of it. This is confirmed by the fact that no science – practical, productive or theoretical – troubles itself about it.[13]

Now metaphysics is one of the theoretical sciences, along with physics and mathematics. And in all case knowledge of what cannot be otherwise is the stated target. In Book 1, chapter 2 of the *Posterior Analytics* Aristotle writes:

> We suppose ourselves to possess unqualified scientific knowledge of a thing, as opposed to knowing it in an accidental way in which the sophist knows, when we think we know the cause on which the fact depends, as the cause of that fact and no other, and, further, that the fact could not be otherwise than it is. (1941, p. 111)

In Book 1, chapter 4 he continues:

> Since the object of pure scientific knowledge cannot be other than it is, the truth obtained by demonstrative knowledge will be necessary. (1941, p. 115)

These points are echoed elsewhere in the Aristotelian corpus. In the *Physics* Aristotle makes it plain that modalities are central to the intellectual task at hand:

> We must explain then...about the necessary and its place in physical problems, for all writers ascribe things to this cause [viz. necessity], arguing that since the hot and the cold, &c., are of such

and such a kind, therefore certain things necessarily are and come to be. (1941, p. 249)

Why focus on those features of reality that cannot be otherwise? It would appear that there are two related reasons. The first has to do with our standing desire for *explanatory* knowledge, the thought being that one has explanatory knowledge of p if one knows not just that p happens to be the case but why p could not have failed to be the case. Now the conceptual link between necessity and explanation is established via the notion of causation, for to know why p could not fail to be the case is to know the causes responsible for p, for a cause, as Aquinas puts it in his commentary on the *Metaphysics*, is simply 'that on which something else follows of necessity' (1995, p. 277). Little wonder then that Aristotle stresses the search for ultimate causes and principles in the opening chapters of the *Metaphysics*.

A second reason for the focus on those features of reality that cannot be otherwise is that any necessity tracked in reality confers a particularly high degree of *certainty* on the beliefs that have tracked that necessity. For one can be particularly *certain* of the truth of the proposition 'X is Y' if one believes of X that it cannot fail to be Y given the way reality is structured. As Aristotle says, *'unqualified* scientific knowledge' is of what is necessary. Now the dependence of certainty on the part of the knower on the objective necessity of the known object is already explicit in the passages quoted from Aristotle earlier. But the important point for present purposes is that this assumption regarding the connection between the modal nature of the reality tracked and the epistemic standing of the tracking belief is one that crosses the boundary between Scholastic and Early Modern metaphysics, and is carried much further, becoming integral to Kant's conceptual framework as well. What changes is where the emphasis is placed, and where the requisite necessity is to be found. As we shall see, Aristotle and the early and middle Scholastics tend to emphasise the necessity of the thing known (rather than the certainty of the knower), and expect to find necessity in the very fabric of things. The late Scholastics and Early Moderns, on the other hand, tend to focus on the certainty of the knower, and expect to find necessity in logic alone (if at all). But the connection between necessity and certainty is part of a shared conceptual framework.[14]

Given the importance of the point for what follows it is worth pausing to providing some textual evidence for this. Aquinas is quite explicit about his following Aristotle on the formal object of metaphysics. He says that the three theoretical sciences, viz, physics, mathematics and metaphysics, while having different formal objects, all have knowledge of necessity as their goal:

> ...for science treats of necessary matters, as is shown in the *Posterior Analytics*.[15]

And in his commentary on the *Metaphysics* he writes:

> [Aristotle] establishes...that there is no science of the accidental. He says that this is evident from the fact that every science is concerned with what is either always or for the most part. Therefore, since the accidental occurs neither always or for the most part, there will be no science of it. (1995, p. 410)

And on the connection between necessity and certainty he writes:

> ...those things from which the intellect derives certitude seem to be more intelligible. (1995, p. xxix)

The first point to notice here is that certainty depends on the nature of the thing known, in this case, the intelligibility of the thing known. But as the intelligibility of X was deemed to be inversely proportional to X's capacity for change (the unchanging being fully intelligible, the always changing being unintelligible), and as a capacity for change involves being able to be otherwise, certainty is possible only of those objects or states of affairs that are not able to be otherwise.

Henry of Ghent is a particularly interesting source on this matter. A reactionary Augustinian, and so not inclined to accept Aristotle's authority uncritically even in philosophical matters, Henry nicely illustrates how natural it was to associate necessity on the part of a known object with certainty on the part of a knower. In what can be called his meta-theology (Articles 6 to 10 of the *Summa of Ordinary Questions*) Henry forwards the daring thesis that of all the sciences theology 'is the most certain'. He says '...the certitude of knowledge is caused by...the truth of the thing known'. Then, relying explicitly on *Metaphysics* 993-b30, which reads '...the principles of eternal things [i.e., unchanging things] must always be most true (for they are not merely sometimes true)', he concludes that certitude is greatest of eternal things. But as God is the eternal truth par excellence, theology is the science that provides the greatest certainty (*Summa of Ordinary Questions*, a. 7, q. 2).

Interestingly Henry does not simply conflate the necessity of a feature of reality with the subjective certainty of a knower with respect to that feature. In an extended discussion comparing the relative degrees of certainty to be attained in the theoretical sciences, he goes on to distinguish three types of certainty. In order to defend the claim that theology is the most certain of the sciences against obvious objections, he distinguishes between the 'certitude of the evidence', the certitude of the 'adherence on the part of the believer' and the 'certitude of the reality' (*Summa of Ordinary Questions*, a. 7, q. 2, p. 63). The point of these distinctions is to allow Henry to acknowledge that the results of theology are less certain than the results of the other sciences in the sense that the evidence for theological claims is less conclusive than is often the case in the other sciences. Nonetheless, Henry maintains that '...concerning the certitude of the reality and of adherence on the part of the believer, the knowledge of the faith can very well be more certain' than the results of the other sciences (ibid., p. 63). It is worth noting that Henry is willing to concede the 'certitude of the evidence' to the other sciences if he is able to retain the 'certitude of the reality' for theology. But most important for present purposes is the phrase 'certitude of the reality'; this is not a phrase we find natural anymore, for it looks like a category mistake. Certitude is a property of a believer, and perhaps by extension it characterises the evidence in favour of a belief. But the phrase is intelligible if we recall the assumed dependence of subjective certainty on the necessity of the thing known.

Similar views can be found in Henry's fellow Franciscans. Scotus, again no passive follower of Aquinas, adheres to standard Scholastic practice and insists that demonstrative proofs require not just true premises and a valid form (as we might expect), but also premises that are necessary in virtue of the nature of the items mentioned in the premises. This commitment is on display in his *De Primo Principio* at the outset of his proof for the three-fold primacy of the first principle. He writes:

> In this conclusion and certain following ones, I could propose the actual thus: Some nature is an efficient cause, because some nature is effected, because some nature begins to be, because some nature is the term of motion and contingent. But I prefer to submit conclusions and premises about the possible...

The reason for this preference is that premises

> ...about the actual are contingent, though manifest enough; those about the possible are necessary. (Scotus, 1949, p. 41)

Here Scotus is merely repeating the line developed by Aristotle in the *Posterior Analytics*. Again, the point is stressed that there can be no science of, and *a fortiori* no certainty regarding, the accidental.

Ockham, so often a critic of both Aquinas and Scotus, follows suit. His discussion of the theological puzzle concerning God's knowledge of future contingents takes as a working assumption the claim that *scientia* is not possible of what is not necessary. He writes:

> It is difficult, however, to see how He knows this [with certainty], since one part [of the contradiction] is no more determined to truth than the other... Secondly, when something is determined contingently, so that it is still possible that it is not determined and it is possible that it was never determined, then one cannot have certain and infallible cognition based on such a determination. (1983, pp. 48–49)

It is easy to find passages like these from numerous Scholastics. But the crucial point here is that similar passages can be found throughout the Early Modern period. Consider Galileo's comparisons of human and divine knowledge which he puts in the mouth of his spokesman Salviati:

> ...I say the human intellect does understand some of these [propositions about the natural world] perfectly, and thus in these it has as much absolute certainty as Nature itself has... With regard to

those few [propositions] which the human intellect does understand, I believe that its knowledge equals the divine in objective certainty, for here it succeeds in understanding *necessity*, beyond which there can be no greater sureness. (2001, emphasis added)

Again we see the link between necessity and certainty at work. The only remarkable point is that Galileo is suggesting that human knowledge can equal divine knowledge in the degree of certainty attained – despite the difference in our respective cognitive capacities – because it is the necessity of the known object that is ultimately decisive.

In the case of Descartes the emphasis is always placed on the achievement of certainty; but it is not difficult to see that for Descartes the dependence of certainty upon necessity remains. This is particularly obvious in his own analysis of the Cogito. He writes:

I observed that there is nothing at all in the proposition 'I am thinking, therefore I exist' to assure me that I am speaking the truth, *except* that I see very clearly that in order to think it is *necessary* to exist. (1990, p. 127)

Here certainty on the part of the knower is grounded in necessity found in the nature of the thing known. And this link is at play in all deductions. Descartes writes:

Deduction...remains as our sole means of compounding things in a way that enables us to be certain of their truth. Yet even with deduction there can be many drawbacks...But it is within our power to avoid this error, *viz.*, by never conjoining things unless we intuit that the conjunction of one with the other is wholly *necessary*, as we do for example when we deduce that nothing which lacks extension can have shape, on the grounds that there is a *necessary* connection between shape and extension. (1990, p. 48)

And in the following passage in a letter to Mersenne we see the same thoughts at work in his physics. Descartes writes:

As to physics, I should think I knew nothing about it if I could only say how things may be without demonstrating that they cannot be otherwise. (in Duhem, 1977, p. 46)

A further important point to note for the purposes of this chapter is that accounting for the possibility of knowledge of what is necessary lies behind the drive to rationalism in the Early Modern period. In his *New Essays on Human Understanding* Leibniz writes:

> The senses...are not sufficient to give us the whole of [knowledge], since the senses never give us anything but instances, that is to say particular or individual truths. Now all the instances which confirm a general truth, however numerous they may be, are not sufficient to establish the universal *necessity* of the same truth....From which it appears that necessary truths...must have principles whose proof does not depend on instances, nor consequently on the testimony of the senses....Logic, together with metaphysics...are full of such truths; consequently proof of them can only arise from inner principles, which are called innate. (1983, pp. 150–151)

Now the point to note for present purposes is that Leibniz's adoption of rationalist epistemological principles is motivated by the concern to preserve and account for the possibility of knowledge of necessary truths. If knowledge of contingent truths were the priority, there would have been no need to bring in innate principles. Confirmation that knowledge of necessary truths is the real prize is implicit in the following lines:

> It is in this also that the knowledge of men differs from that of the brutes: the latter are purely empirical, and guide themselves solely by particular instances; for, as far as we can judge, they never go so far as to form *necessary* propositions; whereas men are capable of the demonstrative sciences. (1983, p. 151)

Obviously Leibniz is not satisfied with knowledge that differs in no important respect from that of the brutes, and he does not expect anyone to demure on this point. And again, the link between necessity and the ideal form of scientific knowledge is clearly on display.

Finally, in Kant we find a similar story. First, that certainty is absolutely essential to his project in the first critique:

As regards the form of our enquiry, *certainty* and *clearness* are two essential requirements, rightly to be extracted from anyone who ventures upon so delicate an undertaking. (1965, p. 11)

And secondly, that certainty is inextricably linked to necessity via the a priori:

As to certainty, I have prescribed to myself the maxim, that in this kind of investigation it is no wise permissible to hold *opinions*. Everything, therefore, which bears any manner of resemblance to an hypothesis is to be treated as contraband; it is not to be put up for sale even at the lowest price, but forthwith confiscated, immediately upon detection. Any knowledge that professes to hold *a priori* lays claim to be regarded as absolutely necessary. This applies still more to any *determination* of all pure *a priori* knowledge, since such determination has to serve as the measure, and therefore as the [supreme] example, of all apodeictic (philosophical) certainty. (1965, p. 11)

These passages are intended to illustrate the shared commitment to the assumption that certainty is intimately connected to necessity, and that the knowledge sought by metaphysicians Ancient and Modern was to display both features. As we shall see, changes in one's beliefs regarding which features of reality can or cannot be otherwise will have profound consequences for one's views on the possibility of certainty, and for one's approach to metaphysics.

A brief history of modality

As we have seen, Aristotle and those who followed repeatedly insist that strict scientific knowledge is knowledge of what is necessary. The point we must now draw attention to is that Aristotle did not believe that scientific knowledge was confined to logic or mathematics, but was possible both in metaphysics and the natural sciences. And in this respect Aristotle thought he was simply following his predecessors. And indeed it would appear that in this respect Aristotle was right about his ancient compatriots. For example, Heraclitus maintained that '...all things take place in accordance with strife and necessity', while Democritus held that 'Nothing occurs at random, but everything occurs for a reason and by necessity'.[16] Talk of 'inescapable and merciless necessity' – Epicurus' phrase – is also found in Plato (1987, *Timaeus*, 47e).[17] Such attitudes found their way into Roman thinking as well. Cicero's report in his *On Fate* clearly implies that it was only in the sphere of human action that it ever occurred to anyone to doubt that the natural world was governed by necessity (Inwood and Gerson, 1997, p. 186).

Things are more complicated in the Middle Ages. But here too we find Aquinas insisting that the natural order has diverse modes of non-logical necessity to be found within it. In the *Summa Contra Gentiles* (Book II, chapter 30) Aquinas sets out the kinds of *absolute necessity* to be found in all created things. This passage is worth quoting in full:

> First, through the relation of a thing's principles to its act of being. Since matter is by nature a being in potentiality, and since that which can be can also not be, it follows that certain things, in relation to their matter, are necessarily corruptible...
>
> Secondly, from essential principles of things absolute necessity arises in them from the order of their parts of their matter or of their form... For, since a human being's proper matter is a mixed body, having a certain temperament and endowed with organs, it is absolutely necessary that a human being have in himself each of the elements and humours and principal organs. Even so, if a human being is a rational moral animal, and this is his nature or form, then it is necessary for him to be both animal and rational.
>
> Thirdly, there is absolute necessity in things from the order of their essential principles to the properties flowing from their matter and form; a saw, because it is made of iron, must be hard; and a human being is necessarily capable of learning.

But the kinds of necessity found in all created things are not exhausted by this list, for Aquinas goes on to spell out in some detail how causes necessitate their effects. This depends on the powers and liabilities of the natures involved in the causal relation:

> Now, the necessity in the effect or thing moved, resulting from the efficient or moving cause, depends not only on the efficient

cause but also on the condition of the thing moved and of the recipient of the agent's action; for the recipient is either in no way receptive of the effect of such action – as wool to be made into a saw – or else its receptivity is impeded by contrary agents or by contrary dispositions in the movable or by contrary forms, to such an extent that the agent's power is ineffective; a feeble heat will not melt iron. In order for the effect to follow, it is therefore necessary that receptivity exist in the recipient, and that the recipient be under the domination of the agent, so that the latter can transform it to a contrary disposition. And if the effect in the patient resulting from the agent's victory over it is contrary to the natural disposition of the recipient, then there will be necessity by way of violence, as when a stone is thrown upwards. But if the effect is not contrary to the natural disposition of its subject, there will be necessity not of violence but of natural order.[18]

The important point here is that Aquinas still believes that metaphysicians can study the fundamental structure of the world insofar as it cannot be otherwise; he still believes that the necessity being tracked is one of the natural order; and that this is possible because the natural order still has the requisite modal features. However, all this is about to change.

The main task of the Scholastic theologian was to produce a coherent body of doctrine based on the narratives of the Old and New Testaments. The inconsistencies and tensions to be found within sacred scripture, as well as challenges to ordinary common sense, are the problems falling to the theologian *qua* theologian.[19] Among Augustine's many historically significant contributions to this effort was his claim that it is permissible for theologians to make use of the philosophies of the pagans in the pursuance of the strictly theological project. This led to the serious and sustained effort on the part of the Scholastics to engage with and employ the work of Aristotle once his entire corpus had been recovered. In the hands of Aquinas and other Scholastics this developed into the concerted effort to synthesise the theology of Augustine with the philosophy of Aristotle.

This project was not without its successes; but fundamental tensions between Aristotelian and Catholic teaching could not be papered over. This led to factionalism within the university system, with traditional theologians on the one hand pitted against the champions of Aristotle on the other, with exchanges becoming increasingly acrimonious. Matters eventually came to a head in 1277 when the Bishop of Paris issued a list of 219 condemned propositions, i.e., propositions that could not be taught within the university on pain of excommunication. The condemned propositions of particular interest for present purposes were those which threatened God's omnipotence. The following is a representative sample:

(1) That what is impossible absolutely speaking cannot be brought about by God or by another agent. This is erroneous if we mean what is impossible according to nature.
(2) That God cannot be the cause of a new-made thing and cannot produce anything new.
(3) That God cannot move anything irregularly, that is, in a manner other than that in which he does, because there is no diversity of will in Him.
(4) That God cannot multiply individuals of the same species without matter.
(5) That God could not make several intelligences of the same species because intelligences do not have matter.
(6) That God could not move heaven in a straight line, the reason being that He would then leave a vacuum.
(7) That God cannot produce the effect of a secondary cause without the secondary cause itself.[20]

These propositions had been entertained by various thinkers taking Aristotle as their point of departure, and all posit a limit on God's power. In the wake of the condemnations such thoughts were no longer tolerated. And as on standard interpretations of omnipotence God's power is limited only by the principle of non-contradiction,[21] it was incumbent upon all to accept as *physically* and *metaphysically* possible any proposition that did not contain or entail a *logical* contradiction. As Nicolas Autrecourt would put it 60 years later, and over 200 years before Descartes: 'Every being which does not contain an incompatibility in its concept is possible' (1971, 42).[22] The upshot of the condemnations for philosophy were: (1) that the notions of metaphysical and physical necessity were ruled out as doctrinally unsound; and (2) as strict scientific knowledge is only of what is necessary, knowledge is only possible of logically necessary truths.

The implications for traditional metaphysics and epistemology became clear in due course.

Soon after the condemnations Scholastics begin operating with a completely new mindset in metaphysics and epistemology, citing God's omnipotence in justification. Moreover, they begin to formulate theses commonly associated with philosophers from the Early Modern period. Some clear-cut examples are worth noticing in this regard. If one compares the views of William Ockham (b. circa 1285, d. 1347) and Nicolas Autrecourt (b. circa 1300, d. after 1347) with the views of Aristotle on the one hand and Descartes on the other, the considerable distance travelled in the direction of modernity is obvious. Consider the central principles of the 'Ockhamist' world view, many of which are simply adopted from the condemnations:

(1) All things are possible for God, save such as involve a logical contradiction.[23]
(2) Whatever God produces by means of secondary causes he can produce and conserve immediately without their aid.[24]
(3) God can save, conserve and produce every reality, be it substance or accident, apart from any other reality.[25]
(4) We must not affirm that a proposition is true, or that something is necessarily required for the explanation of an effect, if we are not led to this by a reasoning proceeding either from a truth known by itself or from an experience that is certain.[26]
(5) Everything that is real, and different from God, is contingent to the core of its being.[27]

The contrast here with the views expressed by Aquinas about absolute necessity being found in created things could hardly be more pronounced. And the consequences of this change of heart were not lost on contemporary thinkers. Consider the views expressed by Nicolas Autrecourt in his first and second letters to Bernard of Arezzo in which he draws out the consequences of a post-1277 environment for Aristotelian physics and metaphysics.[28] He writes:

I propose this inference: 'It is possible, without any contradiction following therefrom, that it will appear to you to be the case, and yet it will not be so. Therefore you will not be evidently certain that it is the case.' (in Bosley and Tweedale, 2006, p. 454)

Here Nicolas has raised the bar to genuine knowledge in a fashion that Descartes himself would have accepted. Only what cannot possibly be false can be accepted as true, and only what is *logically* necessary can never be false. So one can know only logically necessary truths. The road to sceptical conclusions is then made easy – one simply has to show that the contrary of a proposition is logically possible to establish that the proposition cannot be known with certainty. Nicolas puts this principle to work, claiming that

> From the fact that some thing is known to be, it cannot be inferred evidently, by evidentness reduced to the first principle [i.e., the principle of non-contradiction], that there is some other thing. (ibid., p. 455)

This is a view one now associates most readily with Hume, and Nicolas's reasoning is essentially the same as the Scot's:

> In such an inference in which from one thing another thing would be inferred, the consequent would not be factually identical with the antecedent, nor with part of what is signified by the antecedent. It therefore follows that such an inference would not be evidently known with the aforesaid evidentness of the first principle. (ibid., p. 455)

In a particularly revealing example of this mindset at work, Nicolas argues that one cannot know with certainty that individual substances underlie observed properties. He writes:

> When a log or a stone has been pointed out, it will be most clearly deduced that a substance is there, from a belief accepted simultaneously. But this cannot be inferred from a simultaneous belief evidently. For, even if all kinds of things are perceived prior to such discursive thought, it can happen, by some power, namely the divine, that no substance is there. Therefore in the natural light it is not evidently inferred from these appearances that a substance is there. (ibid., p. 457)[29]

Having thus denied Aristotle the right to claim that properties require an underlying subject, Nicolas also shows that one cannot be sure the external world matches our sensory experiences in any fashion whatsoever:

> Every impression we have of the existence of objects outside our minds can be false, since...it can exist, whether or not the object is.

The upshot of his reflections is that

> ...Aristotle in his entire natural philosophy and metaphysics possessed such certainty of scarcely two conclusions, and perhaps not even one...I have an argument that I am unable to refute, to prove that he did not even possess probable knowledge. (ibid., p. 457)

In fact, according to Nicolas, the only substance Aristotle ever had any evident knowledge of was his own soul, precisely the point Descartes would make nearly three centuries later in the Second Meditation.[30]

Now there is good reason to think that Ockham, Autrecourt and others working along the same lines managed to influence latter generations of thinkers, if not directly, then at least indirectly through the work of John Buridan. Buridan (1300–1358) was a nominalist in the tradition of Ockham, and he discussed Autrecourt's arguments in works widely disseminated across Europe which formed an important part of the curriculum in many universities well into the sixteenth century.[31] For example, in *Quaestiones in Aristotelis Metaphysicam*, bk. 2, q. 1, where Buridan discusses whether scientific knowledge is possible, one finds a series of sceptical arguments that are strikingly reminiscent of Descartes' First Meditation and much of Hume as well, and these arguments turn on points insisted upon by Ockham and Autrecourt. There are traditional arguments against the reliability of the senses which Buridan then claims are made even more difficult by articles of the faith, namely, that God can cause us to have visual experiences of objects which do not exist because this is consistent with divine omnipotence. He writes:

> ...for God can form in our senses the species of sensible things without these sensible things, and can preserve them for a long time, and then...we would judge those sensible things to be present.

Furthermore, you do not know whether God, who can do such and even greater things, wants to do so. Hence, you do not have certitude and evidentness about whether you are awake and there are people in front of you, or you are asleep, for in your sleep God could make sensible species just as clear as, or even a hundred times clearer than, those that sensible objects can produce; and so you would formally judge that there are sensible things in front of you, just as you do now. Therefore, since you know nothing about the will of God, you cannot be certain about anything. (in Klima, 2007, p. 143)

It is easy to see this passage as an early version of the infamous malin génie hypothesis of the First Meditation.[32] One also finds an early version of Hume's problem of induction:

Experiences do not have the force to conclude a universal principle, unless by means of induction over many [singular cases]. But a universal proposition never follows by induction, unless induction covers all singular cases of that universal, which is impossible. Indeed, let us assume that whenever you touched fire, you always felt it to be hot; therefore, by experience you judge the next fire, which you have never touched, to be hot too, etc., and so finally you judge that every fire is hot. Let us assume, then, by God's will whenever you touched a piece of iron, then you felt it to be hot. It is clear by parity of reasoning that when you next see a piece of iron that is in fact cold, you will judge it to be hot. And these would be false judgements, although at that point you would have just as much experience about iron as you now in fact have about fire. (ibid., p. 144)

Finally, it is worth noting that Autrecourt's point about the inadmissibility of inferring causes from effects is given an airing in Buridan's discussion:

Again, a conclusion or an effect cannot be known through it cause, or a cause through its effect, because the cause is not contained essentially or virtually in its effect... [So] it seems that we can never have evident knowledge of one thing through another, because there is no evidentness, except by reduction to the first principle, which is grounded in contradiction. However, we can never have a contradiction concerning two diverse things: for let us assume that they are A and B; then it is not a contradiction that A exists and B does not exist, or that A is white and B is not white. Therefore, there will never be an evident inference concluding that A exists from the fact that B exists, and so on for other cases. (ibid., p. 144)

This is simply a repetition of Autrecourt's fundamental point on which he bases his critique of Aristotle's claims to know anything at all. But the point to note here is that the views of Ockham and Autrecourt did get an airing well into the sixteenth century and beyond, views that gained widespread currency only after the condemnations of 1277.

A further point of interest: Buridan *rehearsed* these arguments, but he did not accept their sceptical conclusions. Buridan agreed that acceptance of God's omnipotence does indeed lead to the consequences Autrecourt so ably drew. But he maintained that the philosopher could afford to ignore these results. He writes:

It is in this way [i.e., on the assumption that things obey the common course of nature] that it is evident to us that every fire is hot or that the heavens are moving, although the opposite is possible by God's power. *And this sort of evidentness is sufficient for the principles and conclusions of natural science.* ... (ibid., pp. 145–146)

Indeed he rebukes those who do not settle for this degree of evidentness:

Therefore we conclude as a corollary that some people speak very wrongly, wanting to destroy the natural and moral sciences on the grounds that their principles and conclusions are often not absolutely evident, but can be falsified by *supernatural possibilities.* Because such sciences do not require absolute evidentness, but the above mentioned kinds of non-absolute evidence or conditional evidentness suffices. (ibid., p. 146)

Apparently Buridan did not feel the need to meet sceptical arguments based on supernatural possibilities. They might command the attention of *theologians*, he implies, but they need not detain the *natural philosopher*.

A further link between the Condemnations and Descartes in particular is worth mentioning. Descartes does not betray any direct acquaintance with the works of Ockham, Autrecourt or Buridan, but 'the first philosophical author [he] came across', viz., Suarez, was certainly familiar with this Scholastic heritage.[33] And in the course of his *Metaphysical Disputations*, Suarez relies on the cluster of assumptions whose history we have been tracing. For example, in Disputation 31, section 6, we see the necessity claim at work in the blurring of the distinction between logical and metaphysical necessity: ' ...a thing capable of being created implies merely the lack of contradiction, or logical possibility' (1983, p. 95). And when explicating the difference between modal and real distinctions Suarez relies again on the powers of God to ground the difference. If two putative entities are only modally distinct they cannot be separated even by God's infinite power, which, again, is limited only by the principle of non-contradiction:

> The reason is that if one of these two extremes is of such a nature that this cannot be conserved without the other even by God's absolute power, this is a strong indication that it is essentially no more than some sort of mode rather than a true entity; if it were a true entity, it could not have so intrinsic a dependence on another entity that God could not supply for the dependence by His infinite power. (2007, p. 45)

Just how well Descartes knew the works of Suarez is a matter of some debate; but there is no doubt that Suarez deeply influenced Descartes' teachers at La Fleche, and so a line of influence can be traced from the Condemnations to Descartes.[34] What is not so clear is why Descartes did *not* follow in the footsteps of Buridan, but instead adopted the theologically motivated course taken by Ockham and Autrecourt.[35] But whatever the answer to that question might be, by insisting on absolute certainty, and by assuming that the only necessity is logical necessity, Descartes imported into the natural sciences, metaphysics and epistemology assumptions more naturally at home in the theology of the thirteenth and fourteenth centuries, assumptions with important methodological implications.

The meta-metaphysical and methodological implications of 1277

There are two connected methodological implications of the condemnations of 1277 that are worth close attention. The first, and most obvious perhaps, is that metaphysicians are going to be drawn to mathematics, for here at least one can still find the necessity required to ground the possibility of *scientia*. It is thus not surprising that Descartes, Spinoza and Leibniz all show marked enthusiasm for the methods associated with Euclidian geometry, and that they seek to appropriate such methods for philosophy itself. The idea is that one's metaphysical reflection must take on the form of an axiomatic system of deductions, because this is the only way one can be sure that the objects of metaphysical knowledge have the requisite modal character. The rationalism found in this trio of metaphysicians, the fruits of which are presented most clearly in Spinoza's *Principles of Cartesian Philosophy* as well as the *Ethics*, is the natural result of accepting the changes brought about by the condemnations.[36]

But a second point is that the quasi-mathematical approach to metaphysics characteristic of the rationalists is accompanied by the notion that metaphysical reflection can be carried out 'from the armchair' by engaging in thought experiments that trade on the connection between the conceivable and the logically possible. For if the order of extra-mental reality is a purely logical order, as is required if God's omnipotence is to be respected, it is but a small step to the claim that one can discern the metaphysical structure or order of reality by a priori reflection on the logical order of one's conceptual framework. This is the great methodological consequence of the Condemnations, and it is by far the most important for contemporary metaphysics. For while few metaphysicians today are tempted to emulate the Early Modern rationalists, many still rely on conceivability arguments.

Crucial in this regard is the connection between conceivability and logical possibility, a topic of much interest in contemporary circles. In its most straightforward formulation, the connection in the thesis is that what one can conceive of is a good guide to possibility and necessity. If something, or some state of affairs, or some event is deemed conceivable, then one is warranted in asserting its possibility; if not, then not. A recent expression of this line of thought is found in Williamson:

> ...we assert □A when our counterfactual development of the supposition ¬A robustly yields a contradiction; we deny □A when our counterfactual development of ¬A does not robustly yield a contradiction... Similarly... we assert ◊A when our counterfactual development of the supposition A does not robustly yield a contradiction; ... we deny ◊A when our counterfactual development of A robustly yields a contradiction. Thus our fallible imaginative evaluation of counterfactuals has a conceivability test for possibility and an inconceivability test for impossibility built in as fallible special cases. (2007, p. 163)

Of course, just what conceivability amounts to is a contentious matter. Sometimes the conceivable is taken to be the imaginable, and the two terms are often simply stylistic substitutes for one another. Chalmers (2002) has recently tried to offer some help here by distinguishing various kinds of conceivability, but the core suggestion is that something is conceivable if a conceiver, relying on background knowledge, runs a mental 'simulation' of a putative possibility and notices no *contradiction* arising in the course of the simulation. While conceivability on this account is always relative to a conceiver's background information, the general thrust is that conceivability is grounded in logic. A thing, or a state of affairs or an event is conceivable, and therefore possible, if no contradiction is perceived in the positing of the thing, state of affairs or event.

Now, when one combines the claim that the only necessity is logical necessity with the conceivability thesis it is but a small step to an important corollary which can be called the *Principle of Separability*.

> PoS: If one can conceive of X apart from Y because neither is included in the definition of the other, then X and Y are distinct entities and can exist apart from each other no matter how closely connected they might be. Conversely, if one cannot conceive of X apart from Y because one is included in the definition of the other, then X and Y are not distinct entities, and cannot exist apart from each other.

This principle is of some significance because it warrants the following inferences. Assuming the necessity claim, if one can conceive of X apart from Y, then one can conclude that:

(1) Neither can be reduced to the other because X and Y are not identical.
(2) The existence of the one cannot be inferred from the existence of the other because either can exist without the other.
(3) One cannot be explained in terms of the other, again because one can exist without the other, so one cannot be the cause of the other.
(4) Neither can be part of the mind-independent essence of the other because neither is included in the definition of the other.

In a moment we will see that these inference patterns have played a central role in recent metaphysics. But it is worth pausing to note that the first philosopher in the *orthodox* Western cannon to employ the principle of separability was Descartes, and it lies behind his most characteristic theses.[37] His first, and most decisive, intellectual move was to abandon Aristotelian physics with its particular view of material bodies. Aristotelian physics assumed that material bodies have a number of properties that play an explanatory role in the account of physical phenomena. These properties included 'heat', 'cold', 'moisture' and 'dryness'. By contrast, Descartes insists that his physics will include no reference to such properties because they are not part of the essence of material bodies, and so do not figure in the explanation of physical phenomena. His material bodies have only the properties of motion, size, shape and arrangement of parts.[38] His argument for this claim is that he can conceive of material bodies without their having the properties of heat, cold, moisture and dryness, but he cannot conceive of a material body without its being extended. The same point is made again later in *Principles of Philosophy* (Part II, section 11),

Suppose we attend to the idea we have of some body, for example a stone, and leave out everything we know to be non-essential to the nature of body: we will first of all exclude hardness, since if the stone is melted or pulverised it will lose its hardness without thereby ceasing to be a body; next we will exclude colour, since we have seen stones so transparent as to lack colour; next we exclude heaviness, since although fire is extremely light it is still thought of as being corporeal; and finally we will exclude cold and heat and all other such qualities, either because they are not thought of as being in the stone, or because if they change, the stone is not on that account reckoned to have lost its bodily nature. *After all*, we will see that nothing remains *in the idea* of the stone except that it is extended in length, breadth and depth. (ibid., p. 227, emphasis added)

and repeated in rule 12 of *Rules for the Direction of the Mind*. Concerning 'simples', i.e., entities that cannot be analysed into more basic elements, Descartes writes:

...the conjunction between these simple things is either necessary or contingent. The conjunction is necessary when one of them is somehow implied (albeit confusedly) *in the concept* of the other so that we cannot conceive either of them distinctly if we judge them to be separate from each other. It is in this way that shape is conjoined with extension, motion with duration or time, etc., because we cannot conceive of a shape which is completely lacking in extension, or a motion wholly lacking in duration.... (ibid., p. 46)

The point of these passages with respect to Cartesian physics is twofold. First, they exhibit the principle of separability at work. Second, Descartes' decisive break with Aristotelian physics relies on the implementation of this principle.

It is not just in physics that Descartes employs the principle of separability. The arguments of the First Meditation depend on the idea that no proposition can be known unless it is logically impossible for it to be false. This is most clearly seen in the malin génie scenario. The principle of separability prevents one from arguing from one's perceptions to the nature, or even existence, of an external world because there is no conceptual or logical connection between effects and causes that would warrant such an inference. It is, after all, logically possible (because entirely conceivable) that our perceptions are not caused by anything at all, or caused by a malin génie manipulating our minds in some fashion rather than ordinary material objects in the external world stimulating our sensory receptors. Consider also Descartes' argument for the real distinction between mind and body. Descartes argues that his mind cannot be identified with any material body because he can formulate a clear and distinct perception of their distinctness. At bottom what Descartes is claiming is that our concept MIND does not include, nor is included in, our concept BODY, and so one can conceive of one without the other. Thus Descartes' opening moves in physics, epistemology and philosophy of mind all depend upon his use of the principle of separability.

It is worth stepping back from Descartes' characteristic claims for a moment, and pausing to get clear about the fundamental assumptions underlying the Cartesian framework. The following are the chief characteristics of the Cartesian mindset.

(1) Philosophers should accept only what is certain, or what follows from what is certain by accepted rules of inference.
(2) What is certain is indubitable. Indubitable propositions are either (i) reports of the operations of one's own mind (e.g., 'I am currently experiencing a red patch in my visual field', 'I am now thinking', etc.), or (ii) those that could not possibly be false (e.g., those that are necessarily true).
(3) The only necessity is logical, conceptual or semantic necessity.
(4) Conceivability is a good guide to logical necessity and possibility.
(5) What one can and cannot conceive of or imagine is an important source of information on a range of substantive metaphysical and epistemological issues.

Now while some elements of the Cartesian mindset are no longer widely accepted, a brief survey of post-Cartesian philosophy suggests that (3)–(5) are alive and well.[39] The key point is that they provide the framework in which the inference patterns noted earlier strike one as natural, perhaps even compulsory.

We can begin our survey by noting that the necessity claim was generally taken for granted prior to Kripke's *Naming and Necessity*. Wittgenstein, for example, baldly asserts it with no supporting argumentation whatsoever. Bearing in mind that the last lines of the *Tractatus* appear to retract much that had been previously asserted, Wittgenstein states, but does not argue for, the following claims:

(1) Outside of logic all is accident. (6.3)
(2) A necessity for one thing to happen because another has happened does not exist. There is only logical necessity. (6.37)

Consider also Moore's Open Question Argument against naturalism in metaethics, and Chalmers' zombie argument against physicalism in the philosophy of mind, both of which exemplify (i). Although the proper formulation of the argument is a matter of some debate, if one recalls that Moore is interested in the natures of things referred to by means of language rather than in the meanings of words *per se*, then the OQA of it can be reconstructed along these lines:

(1) If the referent of 'good' is identical to the referent of 'desirable' (as a naturalist might claim) then, as everything is necessarily identical to itself, it is necessarily the case that the referent of 'good' is identical to the referent of 'desirable'.
(2) But as the only necessity is logical necessity, 'good' ought to be conceptually analysable as 'desirable'. That is, 'If x is desirable, then x is good' ought to be an analytic truth because it would amount to a 'barren tautology'.
(3) But 'If x is desirable, then x is good' is *not* analytic because it is deemed to be an open question by all competent speakers of English, so
(4) The referents of 'good' and 'desirable' are not identical, contra the naturalist.
(5) Steps (1) through (4) can be reiterated for any other naturalistic analysis of 'good', so
(6) Ethical naturalism is false.

David Chalmers (1996) is perhaps the best-known proponent of the zombie objection to various forms of physicalism in the philosophy of mind. The objection has been around for some time now, but it has recently returned to prominence.[40] Following Jackson and Braddon-Mitchell (2007, pp. 123–124) it can be presented as follows:

(1) We can conceive of a world which is physically identical to ours but which lacks features which ours has. In particular, we can imagine people (zombies) who are physically identical to us but who lack consciousness.
(2) Conceivability is a good guide to logical possibility.
(3) Zombies are logically possible (from 1 and 2).
(4) We are not zombies as we are conscious.
(5) There is a logically possible world which is a physical duplicate of the actual world while being mentally different (from 3 and 4).
(6) Physicalism, i.e., the view that a complete physical description of the world exhausts all that it contains, is false (from 5).

Continuing this quick survey, we find Nagel's (1974) explanatory gap argument exemplifying (iii). Nagel claims that nothing we currently know about the brain, nor anything we can imagine finding out about the brain, could explain qualia and other aspects of consciousness. On the basis of this he has argued that the mind/body problem is insolvable within our current conceptual framework. His argument can be reconstructed as follows:

(1) Causal explanations in the natural sciences have a kind of causal necessity. (Causal explanations explain why a system *has* to be in the state it is in.)
(2) Necessity implies inconceivability of the opposite.
(3) No account of neuronal activity can explain why, given a particular neuronal state, one *has* to be, for example, in pain because one can always *conceive* of a state of affairs in which a system in that particular neuronal state is not in pain (there is no logical contradiction in such a state of affairs).
(4) Thus there can be no adequate scientific explanation of the mental in terms of the physical.
(5) Thus we will not be able to solve the mind/body problem (at least not within our current conceptual scheme).

Quine's (1953) argument against essentialism exemplifies (iv). He argues against essentialism on the following grounds:

(1) If E were essentially true of X (if E were of the essence of X), then 'X is E' would be a necessary proposition.

(2) If 'X is E' is necessary, then 'X is E' is analytic.
(3) But '"X is E" is analytic' is true or false only relative to our concep-
    tions of X.
(4) So, contra essentialism, 'essential' properties are observer rela-
    tive, not features of things in and of themselves.

These examples of classic arguments from twentieth-century analytic philosophy show that crucial elements of the Cartesian mindset are still alive and well. But I want to end this survey by looking at Hume and Kant, for they are the ultimate source of contemporary concerns about the possibility of metaphysics.

Consider Hume's views on the nature of causation, which still command respect in certain circles, and which exemplify (ii). The line of thought he employs in this context is also that which under-girds his general critique of metaphysics. One historically effective reading of Hume has him submitting the common sense notion of causation to analysis and finding that one of its key ingredients, namely the *necessity* of the connection between cause and effect, does not survive scrutiny. Hume then suggests that this notion of causa-tion ought to be replaced by the weaker notion of constant conjunc-tion. A cause is constantly conjoined with its effect, but there is no necessary connection between the two as we are wont to suppose, or at least we are not justified in believing that there is. The essential claim for our purposes is as follows:

> Upon the whole, necessity is something, that exists in the mind, not in objects; nor is it possible for us ever to form the most distant idea of it, consider'd as a quality of bodies. (1989, pp. 165–166)

The claim that our concept of necessity is never grounded in the nature of things themselves but only in the relation of ideas is supported by two arguments. The first stresses that we have no impression of necessity arising out of our experience of the natural order; the second stresses that there could be no such necessity because cause and effect are two distinct entities. It is the latter argu-ment which is of interest here. It runs as follows:

> The mind can never possibly find the effect in the supposed cause by the most accurate scrutiny and examination. *For the effect is*

> *totally different from the cause, and consequently can never be found in it... every effect is a distinct event from its cause.* It could not, therefore, be discovered in the cause, and the first invention of conceptions of it... must be entirely *arbitrary.* (emphasis added)

This is important because:

> As all distinct ideas are separable from each other, and as the ideas of cause and effect are evidently distinct, 'twill be easy for us to conceive any object to be non-existent this moment, and existent the next, without conjoining to it the distinct idea of a cause or productive principle. The separation, therefore, of the idea of a cause from that of a beginning of existence, is plainly possible for the *imagination*; and consequently the *actual* separation of these objects is so far possible, that it implies no logical contradiction or absurdity; and is therefore incapable of being refuted by any reasoning from mere ideas without which it is impossible to demonstrate the necessity of a cause.[41]

These passages clearly show the principle of separability at work. But the crucial point for subsequent discussion is that Hume's general critique of metaphysics assumes that the only warrants available for any metaphysical thesis are either direct sensory experiences (in modern garb, our best scientific theories) or a priori reflection on the conceptual connections between ideas. Given the background assumption that the only necessity is logical necessity it quickly follows that synthetic but necessary propositions can never be warranted, and so metaphysics as traditionally conceived is impossible. And these are precisely the considerations that led the logical positivists to reject metaphysics in the twentieth century.

Finally, consider some programmatic remarks from Kant: On the standard reading of Kant his project in the *Critique of Pure Reason* takes as one of its points of departure the assumption that Hume correctly asserted that there is no impression of necessity to be had in our experience of the empirical world and that there is no conceptual connection between causes and effects.[42] But, unlike Hume, Kant was not a sceptic about the possibility of scientific knowledge, which he took to be knowledge of necessary propositions, because he believed that human beings actually possess genuine 'a priori' knowledge in the field of the 'general science of nature' as well as 'pure mathematics' (1933, p. 128). In order to account for this obvious discrepancy Kant argued for a possibility that 'never occurred' to Hume, namely, ' ... that the understanding might itself ... be the author of the experience in which its objects are found' (ibid., p. 127). Kant did *not* maintain that the only necessity is logical necessity – there are, after all, on his account, *synthetic* a priori propositions. But because he maintains that necessity is *not* to be found in the world in and of itself, and so cannot be discovered *a posteriori*, Kant is forced to argue that the ground of this necessity lies in the mind-dependent nature of the phenomenal world. It was this which never occurred to Hume; but it would never have occurred to Kant if he had not accepted Hume's two claims.

Now the point of these examples is to show that much contemporary metaphysics has laboured, and continues to labour, in the shadow of 1277. My suggestion then is that if one wants to know how to do metaphysics free of the influence of 1277, free, that is, of theologically motivated modal assumptions that have little to recommend them in the absence of theological support, one could do a lot worse than consult the masters of the discipline who worked prior to the condemnations.

If one takes this suggestion seriously one quickly finds striking contrasts between the methodological recommendations to be found on the other side of 1277. Perhaps the first thing to note is the absence of arguments trading on the inference patterns associated with the principle of separability. What one can conceive of is *not* taken as evidence regarding extra-mental states of affairs prior to 1277.[43] The moral I draw from this is that metaphysicians ought to be very suspicious of thought experiments in general, and conceivability arguments in particular, because we must reject Hume's 'establishe'd maxim in metaphysics, *That whatever the mind clearly conceives includes the idea of possible existence'* (1989, p. 32). This can be no maxim of ours. In the absence of compelling arguments to the contrary, I will assume that there are forms of non-logical necessity to be found within the fabric of the natural order. I will also assume that the existence of uninstantiated possibilities cannot be established merely by their being conceivable.

Secondly, Aristotle and the early and middle Scholastics do not expect respectable metaphysical claims to be established via deductions that meet the standards appropriate to mathematics. To make such a demand, as one finds in the Early Modern rationalists, Hume and Kant, is to depart radically from Aristotle's explicit teaching in numerous works. In the *Metaphysics* (BK II, chapter 3) he writes:

> The minute accuracy of mathematics is not to be demanded in all cases, but only in the case of things which have no matter.

The point that one must adapt one's method and expectations to the subject matter is repeated in the *Nicomachean Ethics*:

> ...for it is the mark of an educated man to look for precision in each class of things just so far as the nature of the subject admits: it is evidently equally foolish to accept probable reasoning from a mathematician and to demand from a rhetorician scientific proofs. (NE, BK 1, chapter 3, in Aristotle (1941))

And in the *Topics* Aristotle makes it clear that it is *dialectical* rather than deductive reasoning that is appropriate to philosophical studies:

> For the study of the philosophical sciences [dialectic] is useful, because the ability to raise searching difficulties on both sides of a subject will make us detect more easily the truth and error about the several points that arise. It has a further use in relation to the ultimate bases of the principles used in the several sciences. For it is impossible to discuss them at all from the principles proper to the particular sciences in hand, seeing that the principles are the *prius* of everything else: it is through the opinions generally held on the particular points that these have to be discussed, and this task belongs properly, or most appropriately to dialectic: for dialectic is a process of criticism wherein lies the path to the principles of all inquiries. (*Topics*, BK I, chapter 2, in Aristotle (1941))

In short, for Aristotle and the early to middle Scholastics the claims of metaphysics are not established deductively, but dialectically. Dialectically established conclusions are usually no more than probable conjectures. But rather than apologising for the conjectural nature of our conclusions, as we are wont to do, we need to recognise that this is precisely what we ought to expect given the nature of our enterprise. The certainty of the knower has to be sacrificed. But we can still see ourselves as trying to identify the necessary features of the structure of reality if we allow ourselves to operate with notions of non-logical necessity. A methodology respecting these two lessons is the focus of the next chapter.

2
The Aporetic Method and the Defence of Immodest Metaphysics

Introduction

In the last chapter an attempt was made to show how it came about that philosophers lost confidence in the viability of metaphysics. The key claim was that metaphysics as traditionally conceived by Aristotle and the early to middle Scholastics is not on all fours with the metaphysics of the early Moderns, and that concerns about the latter do not apply to the former. However, it remains to be seen whether metaphysics, even as traditionally understood, is at all possible, and, if it is, why philosophers should care. It is the intention of this chapter to provide a positive case for the claim that metaphysics is both possible and indispensible to the philosophical enterprise.

To provide such a case we need to consider the following key questions: Do metaphysical questions have answers that are (i) truth-apt, (ii) non-trivial, (iii) tractable but (iv) not provided by the sciences? For much of the last century these questions received a resounding and (almost) unanimous 'No' from the analytic community. But times have changed. The ongoing revival of interest in metaphysics within the analytic tradition itself is testament to the fact that many are now happy to answer each of these questions in the affirmative. But times have not changed that much. Most philosophers continue to eschew metaphysics, abandoning it to a small group of self-selecting enthusiasts who vigorously till the metaphysical garden in splendid isolation. The result is a curiously distorted picture of the state of metaphysics in the analytic tradition. Those who do engage in metaphysical reflection tend to be confident about its prospects, giving the impression that the analytic tradition has restored the queen of the sciences to rude good health; but a sociologist studying the philosophical community would soon discover that the circle of metaphysicians is small, isolated and viewed with indifference or bemused puzzlement by their philosophical brethren.

That the circle of metaphysicians remains relatively small is not surprising. Most philosophers go about their business with no overt engagement with, or apparent reliance upon, substantial metaphysical claims. And this is just as well, they say, because most remain to be convinced of the very possibility of metaphysics. Grounds for concern on this score arise from the following Humean inspired dilemma: On the one hand it is thought that if metaphysical questions can be answered at all they are answered by the sciences or by conceptual analysis (or by some combination of the two) – and in neither case is there any substantial work for the metaphysician *qua* metaphysician.[1] But if metaphysical questions cannot be answered in either of these two ways, then the questions themselves are taken to be defective in some fashion, or perhaps merely verbal, in which case the only legitimate work to do in the area is to rid oneself of the desire to ask metaphysical questions.[2] Consequently it is hardly surprising that most philosophers maintain that we can – and we must – do without metaphysics.

The main burden of this chapter to show what is wrong with this assessment of metaphysics and its prospects. For if metaphysics is ever to be more than a minority interest – let alone restored to its central position within the discipline – this assessment cannot stand. So I present here a case for the indispensability and viability of what might be called *immodest* metaphysics.[3] The central theses are that metaphysics as traditionally conceived is indispensable to the philosophical enterprise; that many non-trivial metaphysical claims can be justified without being 'simply more science'; and finally that accepted interpretations of mature scientific theory will on occasion have to be overturned on the basis of metaphysical reflection. All three theses are substantial, and all three will raise eyebrows. The first depends on a particular view of the philosophical enterprise to be sketched later. The second, pivotal, thesis requires updating and expanding upon Aristotle's aporetic method, wholly neglected in the recent 'meta-metaphysical' literature. I will argue that focusing on aporia provides a method of discovery and justification of metaphysical claims that neatly bypasses the Humean dilemma. The third thesis emerges as a natural consequence of the first two. But it is sensible to begin with a brief account of the metaphysical project as traditionally conceived, and a catalogue of the standard challenges mooted in the literature regarding its viability.

The metaphysical project and its challenges

As noted in the previous chapter, metaphysics as traditionally conceived is the study of the most fundamental structure of reality. On this conception a completed metaphysics would provide an ontology, an account of the relations that obtain between entities of the various kinds and an account of the 'theatre' in which many entities have their being, namely, space and time. Again as previously noted, metaphysicians traditionally do not suppose this study to be straightforwardly empirical in nature if only because it cannot be assumed without further argument that all that exists exists in space and time. Consequently, it is built into the very conception of traditional metaphysics that there will be a need, at least on occasion, to justify in a non-empirical fashion at least some claims concerning the fundamental structure of reality.

Given this conception of metaphysics it is not surprising that many philosophers have wondered whether it is worth the candle. The following list is representative of the familiar kinds of worries philosophers currently express on this score:

1. Ontological relativists and postmodernists deny that there is a single fundamental structure to reality. If 'reality' is a human construct, and variable from culture to culture, then metaphysics as depicted here is a subject without an object.
2. It has been argued that if metaphysics is to be possible at all it must be reconstrued as the study of the fundamental structure of *human thought* or *language*, not of reality itself, either because reality has no structure in and of itself (see 1 earlier), or because the only structure that is cognitively accessible to us is one imposed by our cognitive apparatus.
3. It has been argued that if there is a mind-independent reality whose structure is cognitively accessible to us (contra 1 and 2), the

nature of that reality is discovered empirically. Consequently, if there is any metaphysical knowledge to be had at all, it is provided by the sciences, not by a distinct study called metaphysics.

4. If it is suggested, contra (3), that the structure of mind-inde-pendent reality is cognitively accessible to us via *non*-empirical means, the response is that non-empirical, or a priori, knowledge of extra-mental reality cannot be squared with the evolutionary history of our species. Epistemological naturalists insist that there is nothing in our evolutionary past to suggest that our species came under selective pressures that would lead to the emergence of the required cognitive faculties. The upshot of this naturalism is that appealing to one's a priori reflections concerning one's 'intui-tions', or a priori reflections on what one regards as conceivable, is methodologically dubious at best when developing ontological theories. But if ontological theories are justified empirically – the obvious alternative – then metaphysics is just more science (see 3 earlier).

Now on the basis of these sorts of considerations a number of increasingly sceptical positions have been maintained. First, it is often concluded that warranted answers to metaphysical questions, when available, are provided not by metaphysicians *qua* metaphysi-cians but by the sciences and conceptual analysis (naturalism). If neither of these approaches offers an answer, two attitudes remain. One might still maintain that metaphysical questions have deter-minate answers but judge that one is never in a position to recog-nise the correct answer when it is lighted upon (scepticism). Or one might begin to suspect that metaphysical questions as traditionally conceived are defective in some fashion, and so lack determinate answers altogether (ontological anti-realism).[4]

This is a powerful set of objections to the viability of the metaphys-ical project as traditionally conceived.[5] But there are quick answers to at least the first two objections. To the ontological relativists and postmodernists who would deny the viability of metaphysics on the grounds that there is no single fundamental structure to reality to be discovered by metaphysical reflection, the quick response is that this denial is itself based on a metaphysical claim (that there is no one fundamental structure to reality), and so the objection is self-defeating. To the neo-Kantian who would have metaphysicians focus on the structure of human thought or language rather than on reality itself, two lines of response are needed depending on the justification given for the proposed retreat to thought and language. If this retreat is recommended on the grounds that there is no fundamental structure to reality in and of itself, then the response is the same as that offered to the ontological relativist and postmodernist – the rejection of metaphysics itself depends on a metaphysical claim. If it is proposed on epistemological grounds – viz., the mind-independent structure of reality is not directly accessible to us, and so certainty is not available in this domain, whereas certainty is available regarding the phenomenal world because its structure is cognitively available to us as we have direct access to it – then two lines of response are available. First, the metaphysician can reply that she is not obliged to maintain that metaphysical theses can be established with anything approaching Cartesian certainty. All that is required is that she be able to show that intellectually respectable adjudication between ontologies is possible, and so belief in an ontological claim can be rationally warranted. Second, she can attack the premise on which the retreat is based. For it would appear that at least some aspect of the structure of mind-independent reality *is* accessible to us, contra the initial assumption, because our cognitive apparatus is itself mind-independent (it is not an artefact, or socially constructed). So there is little here to force the metaphysician to retreat from a study of the structure of reality to a study of the structure of human thought alone.[6]

Objections (3) and (4) are altogether more pressing. But there are reasons for thinking that room must be made for the metaphysician *qua* metaphysician even by those who would subcontract the discipline out to the sciences. Some of the more telling considerations are as follows:

1. The sciences cannot provide answers to metaphysical questions in any straightforward fashion because the sciences do not as yet agree amongst themselves on metaphysical matters. So unless the metaphysician is to wait until the sciences are completed, and are found to be in complete harmony, she will have to decide which scientific theories to accept as the basis of her metaphysical reflections. As it is far from obvious that this decision can be based on purely scientific criteria, she will be forced to employ more than the resources of the sciences in the pursuance of the metaphysical project.[7]

2. Even if the metaphysician agrees to set aside consideration of abstract objects and focus entirely on the denizens of space and time, the assumption that the sciences can determine what actually exists in their respective domains unaided by metaphysical theory is arguably false. Empirical evidence counts in favour of the actual existence of X only if it has been established independently that X is possible. But establishing the possibility of X does not fall within the remit of the sciences. That is the business of metaphysics.[8]

3. Metaphysics, and ontology in particular, can be subcontracted to the sciences only if there are good reasons to believe that mature scientific theories are at least approximately true representations of some aspect of reality. But it is not clear that this attitude vis-à-vis any scientific theory can be warranted in the absence of metaphysical commitments. For, as Duhem pointed out long ago, it is impossible strictly speaking to verify or falsify a scientific theory if one relies solely on empirical evidence and logic. For the logic of falsification is such that whenever a recalcitrant observation arises it is open to the scientist to abandon or qualify the theory being tested, or an auxiliary hypothesis, or to question the initial conditions under which the experiment or observation was carried out – the choice not being forced by the empirical facts or logic alone. Now if the choice to save or reject the theory in question is based on pragmatic considerations alone, then the theory has not been saved or rejected on truth related grounds. Consequently one is not entitled to believe that the theory is approximately true (if it is saved), and so it cannot legitimately be used as the basis of ontological reflection. To save or reject a theory in a manner consistent with its use in ontological reflection, one must have some truth related grounds that allow one to weigh the credentials of (i) the recalcitrant observation(s), (ii) the initial conditions, (iii) the instruments and techniques employed, (iv) the auxiliary hypotheses and (v) the theory being tested. Now it is generally agreed that, in practice, such judgements are guided by what have been called 'disciplinary matrices', complicated and usually implicit agreements reached by the scientific community regarding the assumptions of the discipline.[9] But while the sociological aspects of the disciplinary matrices have been widely recognised and accepted, what has not been emphasised sufficiently is the metaphysical import of many of these assumptions.[10]

The upshot of these arguments for and against the possibility of metaphysics I take to be as follows: The standard objections to metaphysics are not as strong as commonly supposed. Either they are self-defeating (1 & 2), or there are reasons for thinking that metaphysics as traditionally conceived is necessary even for those whose ontological faith is placed in the sciences (3). Nonetheless, the rehabilitation of metaphysics requires a response to the outstanding challenge, namely, showing how metaphysical claims can be warranted without being just more science or relying on a dubious a priori methodology. To meet this challenge a new (or rather, a very old) picture of metaphysics – embracing its rationale, activities and possibilities – is required. I turn then to the task of presenting metaphysics in a different, and hopefully attractive light. I begin with the role of metaphysics within philosophy at large.

The philosophical enterprise

The next step in this defence of metaphysics is to establish that there really is a role for metaphysicians within the wider intellectual economy, and within philosophy in particular. The necessity of metaphysics rests on two claims. First, that reflection of a distinctly metaphysical variety is called for in the treatment of problems of a very specific form. Recognition of this point is crucial if the metaphysical project is to be understood and its claims evaluated appropriately. The second claim is that problems calling for metaphysical reflection are typical of the discipline of philosophy in general, and that to discuss the nature of these problems is to discuss the nature of philosophy itself. If these claims can be made good then it follows that metaphysics is indispensable to the philosopher because metaphysics is the core discipline of philosophy. I begin then with an account of the particular kind of problem that calls for treatment by the metaphysician *qua* metaphysician.

The account of these problems and philosophy in general is based primarily on Aristotle's remarks in the *Topics, Posterior Analytics, Nicomachean* and *Eudaimon Ethics* and *Metaphysics*. But it also builds explicitly on the views of modern philosophers, such as Gilbert Ryle, Wilfred Sellars and Nicolas Rescher. That this account shares features with those offered by thinkers of a variety of different stripes and historical periods suggests that the account offered here has some

staying power. One of its main virtues, I suggest, is that it identifies a legitimate role for philosophy within the general intellectual
economy.[11] It also squares rather well with most of the history of the
discipline, and can accommodate a wide range of meta-philosophical
views. But the principal interest of this account for present purposes
is that philosophical problems are seen to have a characteristic structure which permits resolution in a strictly limited number of ways.
In some cases no explicit appeal to metaphysics will be required for
the resolution of a problem. But in many cases such appeals will be
unavoidable.

I begin then with the general aim of philosophical activity in its
broadest sense. The aim of the discipline throughout most of its
history has been to provide a description and explanatory account
of the nature of reality and the place of human beings within it. The
idea, explicit in some and implicit in others, has been that knowing
something about the nature of the world we live in, and something
of our own human nature, is bound to shed light on what kind of life
human beings should lead and what kinds of actions human beings
ought to perform and which to avoid. It is for this reason that philosophy is always associated with the 'Big Questions'. Implicit in this
picture is the idea that philosophy is not just a theoretical exercise,
but is ultimately connected, if at times somewhat distantly and indirectly, with the practical and existential concerns of ordinary life.[12]

Now the project as outlined earlier is in one important respect
not as grandiose as one might think. And the reason for this is that
philosophers *qua* philosophers do *not* provide the basic materials
out of which 'The Big Picture' is developed. If one is to understand
the distinctive nature of philosophy one must begin by recognising
that a division of intellectual labour exists in the general intellectual economy between philosophy on the one hand and the sciences
and truth-directed subjects of the humanities on the other. It is the
role of the special sciences, for instance, to conduct investigations
into that aspect of reality peculiar to them, and to discover new
facts and develop theories within and about that particular realm
or domain. By contrast, there is no particular aspect or element of
reality that philosophers study *qua* philosophers, as there is, say, for
the biologist, chemist or economist. Nonetheless philosophy is not
open to the complaint that it is a subject without an object, because
philosophy is not a first-order discipline on a par with the special

sciences. Rather the contribution of the philosopher *qua* philosopher to the grand project is to draw on pre-existing materials derived from the special sciences, the truth-directed subjects of the humanities as well as our store of pre-theoretical beliefs, and to *coordinate* this material into a coherent picture of human beings and our place in the Universe. It is this task of coordination, lying outside the remit of any special science, which is specifically philosophical, and the problems encountered in the pursuance of this task are specifically philosophical problems.[13]

A principal thesis of this account of philosophy then is that strictly philosophical questions arise out of coordination problems. It is important therefore to say something further about the nature of coordination problems in general. A coordination problem initially arises when one notices a tension, real or apparent, between beliefs or lines of thought that one is otherwise inclined to accept – when a presupposition or entire line of thought from one special science, for instance, appears to clash with a belief from a distinct science, or theology or common sense. The problem is that each line of thought is attractive and well established within its respective domain, but the taking up of the one precludes, or at least appears to preclude, the taking up of (all) the others. Such tensions are commonly felt to be significant because the conflicting beliefs are usually important elements of The Big Picture it is the aim of philosophy ultimately to generate.

As an illustrative example consider the perennial free will/determinism debate. It arises because a tension is noticed between individually plausible theses:

1. Adult human beings are appropriate subjects of reactive attitudes because they are free agents: we can and sometimes do act from free choice, and so we can be held responsible for our actions (an assumption of pre-theoretical common sense, and our legal system).
2. If an action issues from free choice, then it is causally unconstrained (a natural interpretation of the presuppositions of free action according to common sense).
3. All occurrences, human actions included, are caused by antecedent occurrences (a presupposition of the sciences regarding phenomena above the micro-level).

The problem is that these individually plausible theses appear to be inconsistent. The philosophical task is to determine what to make of this situation. Examples of such puzzles abound, and following Aristotle we can call them 'aporia'.[14] But the crucial point for present purposes is that aporia fall to philosophy, as opposed to a special science, if only because such puzzles usually do not arise *within* the domain of any special science, but are due to a prima facie clash *between* first-order disciplines, or between first-order disciplines and common sense, and thus lie outside the competence of any first-order discipline. But the frequently interdisciplinary nature of such problems points to the deeper fact that it is not first-order empirical work that is required for their resolution (if more empirical information is enough to solve a problem then the problem was not a strictly philosophical problem at all) but a modification of our conceptual scheme with which we interpret the empirical data thrown up by the first-order disciplines.[15]

Now if strictly philosophical problems or questions begin as and emerge from coordination problems as described earlier, then the following can be said about philosophical activity in general at the highest level of abstraction: The task of the philosopher *qua* philosopher is to give an account of the initial lines of thought that removes the appearance of contradiction, and so solves the philosophical puzzle. Removing these tensions constitutes success in philosophy. Indeed, in this view, solving problems of this sort is the *raison d'être* of the philosopher *qua* philosopher, and constitutes the philosopher's specific contribution to the general intellectual economy.[16] For by removing the puzzlement one finds that elements of The Big Picture that previously would not fit obligingly together are now coordinated and allotted their respective places. From this perspective all conceptual analyses, second-order theory construction, argument development, analysis and critique, the careful drawing of distinctions – i.e., all the working philosopher's bread and butter activities – are best understood as means to this end, and they receive the tag 'philosophical' because they can be used to this end.[17]

Now if this is what a strictly philosophical problem looks like, and what success in philosophy consists in, then, again at the highest level of abstraction, solutions to philosophical puzzles can take only a limited number of forms. After due consideration the philosopher must either:

(A) Establish that the tensions in the initial lines of thought are merely *prima facie* (perhaps stemming from certain misunderstandings either of the facts of the case, or of our own conceptual system or ontological commitments)

or,

(B) Accept that the lines of thought form an inconsistent set.

Ideally the philosopher finds that the tensions are merely apparent, for this allows one to integrate the original lines of thought into the Big Picture without further ado.[18] However, few today believe that *all* aporia can be handled in this fashion. Many aporia involve genuine contradictions, and in such cases three less than optimal but unavoidable options remain. If the philosopher is convinced that the lines of thought form an inconsistent set, she can maintain that:

(B1) *No totally satisfying account of the inconsistent set of propositions can be given; nonetheless* none *of the propositions should be abandoned.* At least two versions of (B1) are possible. A philosopher might take this position because she is convinced that we simply do not have enough empirical information to solve the problem at present. The recommendation is that one must live in the hope that evidence will eventually come to light which will show one or more of the initial propositions to be false. In this case the aporia is thrown back to the first-order disciplines as being an empirical and not a philosophical matter. But a philosopher might also take this position on the grounds that the aporia is not solvable even in principle because she maintains that reality itself is not consistent (contradictory propositions can be true simultaneously). This line is open to those willing to admit that reality is not fully intelligible, and that this fact must simply be accepted as a genuine feature of the intellectual landscape. Both versions of (B1) are failures in the sense that neither removes the aporia – albeit for very different, and in some instances, enlightening reasons. But it is also the case that both make substantial metaphysical claims, the first that reality *is* structured in such a way as to ensure its ultimate intelligibility (at least in principle), the second that it is not.

But a philosopher of a different temperament might choose a second way of responding to the aporia. She might claim:

(B2) *No coherent account of the inconsistent set of propositions can be provided, so* all *of the initial beliefs fall under suspicion.* One might take this view because all the lines of thought leading to the aporia are deemed upon reflection to concern domains beyond our cognitive competence. A philosopher might argue that we are prone to systematic error in these particular domains because of some feature of our cognitive apparatus. In these cases the diagnosis is that our willingness to entertain beliefs beyond our cognitive competence is the source of the aporia, and we should refrain from entertaining such beliefs. But one might also maintain that the aporia arises because all the initially attractive lines of thought are committed to a similar false assumption. The suggestion here is that there is no problem in principle with our cognitive abilities in the domains under discussion; it is just that we have gotten off to an avoidable false start. The suggestion is that the removal of the common false assumption leads to the resolution of the aporia. The idea here is that the original propositions that constitute the aporia really were in tension (they could not be true simultaneously) but all were fundamentally misguided in some fashion. Both versions of (B2) allow one to resolve the aporia, but at a very high cost. Both require the rejection of the entire set of what had been deemed epistemically credible lines of thought prior to philosophical reflection.

A third option remains. After convincing herself that the aporia is not merely prima facie, the philosopher might claim that:

(B3) *The aporia is to be resolved by modifying, qualifying or perhaps abandoning altogether one or more of the initial lines of thought.* This approach is most likely to be adopted by the philosopher who cannot bring herself to accept (A), (B1) or (B2). That is, by the philosopher who maintains that not all philosophical problems are the result of conceptual confusions of some sort or another; that philosophers cannot be satisfied in all cases with the issuing of promissory notes; that it is methodologically more fruitful to proceed on the assumption that reality is intelligible until very strong arguments to the contrary are produced; and that it is methodologically more fruitful to proceed on the assumption that the various first-order disciplines are not subject to widespread and systematic error. The task for such a philosopher is then to establish in a principled fashion which of the initial lines of thought is to be saved and which sacrificed.

If space permitted it would be helpful to provide further examples of aporia, and of treatments falling into these various categories just identified. But perhaps enough has been said for present purposes.[19] While confident that the forgoing general account of the philosophical enterprise and the logical structure of philosophical problems does identify something fundamental about the discipline, I would not wish to suggest that every instance of philosophical activity will fit this model exactly. What I would claim for the account however, is that it does capture the *focal* sense of the term 'philosophy' inasmuch as strictly philosophical problems appear to begin life as coordination problems. This does not preclude the possibility that other sorts of problems and activities can rightly be termed 'philosophical'. But, to take another page from Aristotle's copy book, in the same way that one can speak intelligibly of healthy diets, healthy lifestyles and healthy urine because these are among the causes or indications of health within an organism, I would suggest that activities or problems are genuinely philosophical insofar as they emerge in the course of coming to terms with a coordination problem. In short, it is these coordination problems that provide the focal sense of the term 'philosophy'.

Aporia resolution

If this account of philosophy is taken seriously then aporia resolution is the philosopher's principal occupation. But how does one resolve an aporia? As just outlined, there are variations on four general stances one can take vis-à-vis any aporia. But how does one decide in a principled fashion which option is best in any given case?

In line with the respect accorded the first-order disciplines recommended by this understanding of the philosophical enterprise, in the first instance the philosopher ought to privilege the first, compatibilist, option. That is, the philosopher ought to assume that the aporia is merely *prima facie* until this proves untenable. And there is no doubt that aporia do often arise because the theories or lines of thought that lead to the aporia are not properly understood. In these instances an aporia can be resolved simply by doing one's home-work, i.e., by making sure one knows what the theories are really claiming, and what their implications really are. In some cases no specifically philosophical training or methodology will be required in order to resolve them because careful attention to the details of the initial material will suffice. In other cases sophisticated forms of paraphrasing might be employed to show that a theory does or does not have a particular implication, and so does not generate an incon-sistent set of propositions. Furthermore, in a large number of cases aporia arise from the failure to recognise certain distinctions. In such cases the philosopher's knack of drawing the necessary conceptual distinctions proves crucial.[20]

But there is no guarantee that all aporia are merely *prima facie*. In such cases no amount of background work or nice distinctions will help. And if one is not prepared to abandon hope of philo-sophical progress (B1), or the wholesale rejection of the first-order disciplines (B2), then one must try to establish which of the initial lines of thought leading to the aporia is to be dropped (B3). One approach to this task has been developed by Rescher, and it deserves close attention.

Rescher's pragmatism

At the bottom of a philosophical problem is knowing what to make of situations in which individually plausible theses are found to be incompatible. Rescher suggests a three-stage approach. One begins by gathering all the data relevant to the aporia. This can be informa-tion gathered from the senses, memory, cognitive aids and instru-ments (telescopes, calculating machines, reference works, etc.) reliable witnesses and the declarations of experts from any relevant discipline. Rescher's working assumption here is that one should accept this data as reliable 'in the absence of specific indications to the contrary' (2009, p. 17). Second, one draws up an inventory of all the possible conflict-resolving options. The example of the free will debate mentioned above provides an illustration. If one decides that the tension is not merely *prima facie* one is left with the following possibilities:

1. Reject 1 and deny that free choices are possible (determinism).
2. Reject 2 and insist that free actions can have a sufficient set of causes (compatibilism).
3. Reject 3 and insist that some events above the microlevel, including some human choices, are uncaused (libertarianism).

The third and final step in Rescher's approach is where matters become problematic, viz., fixing on one of the identified conflict-removing options. When deciding which of the logically possible solutions to adopt he says one must employ 'guidance of plausibility considerations, subject to the principle of minimizing implausibility' (2009, p. 23). In short, one must 'make the less plausible give way to greater plausibility' (ibid., p. 119). Of course as it stands this amounts to little more than a restatement of the problem, for determining which option is most plausible is precisely the issue at hand.

Rescher's suggested way forward begins with the observation that there are three distinct considerations that come to the fore in aporia resolution contexts. First, we value theses for which there is a great deal of empirical evidence, for such evidence provides a degree of confidence and security – a key cognitive desideratum. Second, we are not simply in the business of collecting facts; we want to understand the facts, and so we value theses with a high degree of explanatory power. Such theses tend to be far more general than individual empirical claims, and they tend to rely on core elements of our basic conceptual scheme. Finally, as the first two desiderata are in tension inasmuch as each comes at the expense of the other, one is also likely to seek a balance between security and explanatory power in a coherent system of beliefs that includes both empirical facts and an explanatory framework.

The question then, according to Rescher, is which of these desiderata to emphasise on any given occasion. Rescher's advice is to consider the context. He writes: 'All in all ... the rationale for a particular mode of prioritization lies in the specific goal and purpose of the domain of deliberation at issue. Just this essentially *pragmatic* consideration must be allowed to determine the correlative principle of prioritization' (ibid., p. 139). But this advice is unsatisfactory for a number of reasons. First, frequently there is no single domain involved in the case of aporia. Indeed it is precisely because various domains are in conflict that the aporia arises in the first place. And these domains are likely to have different goals and purposes. Second, even if there were a single domain at issue, it is not clear that Rescher's advice addresses the problem at hand. The problem was to find a way of distinguishing the more from the less plausible. But identifying a goal or purpose for a domain does not allow one to make this adjudication. So while Rescher has a very clear view of the nature of the philosophical dilemma, he has yet to provide a principled way of making the less plausible give way to greater plausibility because the pragmatic method never really addresses the issue of relative plausibility at all.

Metaphysics and plausibility

It is here that one can begin to see why metaphysics as traditionally conceived is indispensable to the philosophical enterprise. If the focal philosophical task is aporia resolution, and if that means, in a significant number of cases, making the less plausible give way to greater plausibility, then some way of determining the plausibility of theories or lines of thought is required. Moreover, if both conflicting lines of thought are justified on empirical grounds (as they will be if they are scientific theories) then more empirical evidence is unlikely to help. It is more likely that a reinterpretation of the available empirical evidence is what is required. And this is precisely what a metaphysical theory is meant to offer by suggesting alterations to our account of the fundamental structure of reality and our accompanying conceptual scheme. Moreover, if one has a respectable theory of the fundamental structure of reality then one has something against which the plausibility of other lines of thought from the first-order disciplines can be measured. If a line of thought is compatible with a respectable metaphysics, then it can be deemed plausible; if not, then it is up for reinterpretation or possibly outright rejection.

And it is here that one also begins to see how it can arise that the humble metaphysician could potentially be called upon to correct the august scientist (although both hats can be worn by one and the same person). If two conflicting lines of thought happen to be scientific theories, and the conflict is not merely *prima facie*, then something has to give. And in these cases the solution to the conflict will not be found within either of the conflicting sciences, and neither science can claim the authority to overrule the other without basing that claim on some extra-scientific judgements. Some appeal to plausibility considerations ultimately grounded in a metaphysical theory will be required to determine which of the sciences needs to be revisited with a critical eye. Thus the eye-catching immodesty of metaphysics as traditionally conceived stems ultimately from the fact that sciences can and do disagree amongst themselves, and because scientific considerations alone will not resolve conflicts of this sort.[21]

The aporetic method

I have now begun to sketch a picture of metaphysics which renders plausible the view that it is indispensable to the philosophical enterprise. And this picture makes sense of the idea that mere metaphysics could reasonably be called upon to overrule an accepted interpretation of scientific theory. It is now time to address perhaps the most pressing concern: providing a warrant for metaphysical claims. For even if metaphysics is necessary, our confidence in a recommended resolution of an aporia will only be as great as our confidence in the metaphysics on which it is based. And metaphysical theories and ontologies abound. How is one to choose in a principled fashion between them? In particular, we need some way of determining when an ontological or metaphysical claim is rationally warranted which (1) does not rely on empirical evidence alone (as this leaves one open to the 'just more science' objection); (2) does not rely on ultimately ungrounded intuitions (however compelling they may appear from the armchair); but (3) is consistent with what we know of the phylogeny of our cognitive systems. I believe it is here that the aporetic method comes into its own.[22]

The first thing to note about the aporetic method is that it starts with aporia, not with a critical discussion of distinctly metaphysical claims *per se*. On this view of metaphysics, metaphysical reflection is only called for in order to resolve independently existing aporia. So the first step for any would be metaphysician is aporia identification. Now the rationale for this crucial first step is not merely that adherence to it keeps metaphysics in touch with the rest of the intellectual economy (a desideratum in and of itself). The method, when properly deployed, guides one's metaphysical research and leads to the discovery of novel theses. Thus at the beginning of Book III of the *Metaphysics* Aristotle states that one must begin one's metaphysical reflections by familiarising oneself with the standard puzzles or aporia. These aporia set the agenda for one's reflections, and one who is not aware of these 'are like those who do not know where they have to go'. Moreover, if one is not familiar with the standard puzzles one will not recognise a solution if one happens to stumble across it – 'a man does not otherwise know even whether he has at any given time found what he is looking for or not; for the end is not clear to such a man, while to him who has first discussed the difficulties it is clear'.

But aporia constitute not just a budget of problems to be solved, but also opportunities for discovery. Aporia are not simply intellectual inconveniences that need to be tidied up. Aporia reveal the cracks and fault lines in our account of reality. These cracks show that our current picture is inadequate (something one might otherwise miss) while also suggesting where the inadequacies lie, thus prompting research in specific directions.

An illustrative example will help. Einstein's discovery of the Special Theory of Relativity appears to have grown out of his reflections on the perceived tensions between lines of thought he accepted.[23] For present purposes we can leave the details as to how these tensions arise to one side and focus only on the logical structure of the problem confronting Einstein. The first line of thought, derived from Newtonian mechanics, asserts that the laws of mechanics take the same form in all inertial frames. The second, derived from Lorentz's interpretation of Maxwell's laws of electricity, magnetism and optics, asserts that there is an inertial frame in which the speed of light is constant regardless of the velocity of its source. The problem is that these two lines of thought lead to a contradiction – namely, that the speed of light is and is not constant across all inertial frames – *if* one also accepts the Newtonian law of the addition of velocities *and* one rejects the ether concept. Now the law of the addition of velocities had always been considered obvious by Einstein and others. And the ether concept, while central to the Lorentz interpretation of Maxwell, had become suspect in Einstein's eyes. He was thus faced with a classic aporia – independently plausible lines of thought from distinct branches of physics are found to be in tension.

Einstein's breakthrough came when it occurred to him to question the apparently obvious law of the addition of velocities (the only option left on the table if one refuses to abandon the other elements of the aporia). He began to consider the implications of abandoning this law, asking what reality would have to be like if this law were in fact false. He realised that rejecting the law would require a new understanding of time, simultaneity and length. With this new analysis in hand Einstein was able to resolve the aporia and maintain that the speed of light is constant in every inertial frame of reference. But the key point for present purposes is that it would probably never have occurred to Einstein to revisit the concepts of space and time in this way if this had not been suggested by the aporia before him. The moral of the story is that aporia can serve to guide one's metaphysical reflections by posing very specific questions of the following general form: If the tensions in an aporia are not merely prima facie, then what must reality be like for these individually warranted but incompatible lines of thought to appear plausible?

The suggestion then is that aporia set the metaphysician to work in the right direction with the reasonable expectation that a suitably able imagination will light upon a better picture of reality, a picture which will explain how a false theory could nonetheless appear compelling, and provide an alternative interpretation of the data leading to the aporia.

But metaphysicians have fertile imaginations, and they often produce very different pictures of reality in response to one and the same aporia. How does one adjudicate between them? The aporetic method itself suggests a test. As the method assumes that the lines of thought leading to the aporia have genuine merit, *the least revisionary of the pictures is the most plausible*. This is not a matter of preserving for preserving's sake. The desire to conserve as many of the original lines of thought as possible stems from the respect accorded to the first-order disciplines implicit in the division of intellectual labour accepted by this account of the philosophical enterprise. For all its immodesty, the approach to metaphysics defended here takes it as a default position that the sciences are competent to run their own affairs, similarly with the truth-directed subjects of the humanities, similarly with the common man's ability to see his way through the tasks of daily life. It is only when tensions between these respectable lines of thought are noticed that there is any call for philosophy or metaphysics. And even then the initial hope is that the tension is merely prima facie. But if it should arise that the tensions are genuine, and a metaphysical theory is required for their resolution, then that theory is judged most plausible which preserves as many of the initial lines of thought as possible because *any* philosophical theory is indirectly supported by the evidence adduced in favour of the initial beliefs it is designed to accommodate. The more such lines of thought are preserved, the greater the degree of support the metaphysical theory enjoys. And if that metaphysical theory should in turn provide the key to the resolution of a range of other seemingly unrelated aporia, then the warrant for that theory becomes all the stronger in virtue of this happy 'consilience of inductions'.

To summarise: On the line taken here, a warrant for a metaphysical claim should take the following schematic form:

Step 1: Establish that T_1-T_n are respectable lines of thought from first-order domains.

Step 2: Establish that T_1-T_n form an inconsistent set of propositions (i.e., that the tensions are not merely prima facie).

Step 3: Construct a theory such that if it were true, then T_1-T_n would appear individually plausible despite the fact that T_3, say, is false.

Step 4: Establish that no other theory is available which preserves as many of the initial lines of thought while explaining how error entered the beliefs of competent authorities.

Step 5: On the basis of steps (1)–(4) conclude that the theory presented in (3) is more plausible than any available alternative.

Step 6: On the basis of step (5) conclude that one is entitled to resolve the initial aporia identified in steps (1) and (2) by rejecting T_3.

Step 7: Test the theory introduced in premise (3) further by considering other aporia. If that theory allows for similarly smooth resolutions of other aporia then that theory increases in plausibility.

Now I submit that this approach to the justification of metaphysical claims meets several key desiderata. First, if one has completed steps 1 and 2 in any particular case then it would appear that there is a job in the intellectual economy that falls to the philosopher, and that in discharging this duty it will be necessary on occasion to appeal to a metaphysical theory. Second, metaphysical theories generated in accordance with these steps appear to be rationally warranted without being just more science, or being grounded in mere a priori 'intuitions', so the Humean dilemma has been circumvented. Finally, there has been no need to postulate cognitive faculties that cannot be squared with our evolutionary past. The aporetic method as sketched here relies on no such dubious faculties, and is entirely consistent with epistemological naturalism. I conclude then, contra current orthodoxy, that immodest metaphysics is both indispensible to the general philosophical enterprise, and a viable project to boot.

Some objections to the aporetic method

It is worth considering briefly some likely objections. The aporetic method as such has not been discussed in the contemporary meta-metaphysical literature, but one obvious criticism is that it is simply a variation on the standard 'battery-of-criteria' approach employed by mainstream metaphysicians.[24] It might reasonably be thought that the aporetic method simply forwards yet another criterion, in this case a theory's ability to preserve more initial lines of thought from first-order domains than any other.

But in reply one can say that this criterion has a claim to being a 'master' criterion, not merely one of a list of competing criteria without a clear hierarchy to be found amongst them. After all, this criterion is tailored precisely to the intellectual task aporetic metaphysicians have set themselves. Here the wider purpose of metaphysical reflection, i.e., aporia resolution, is critical in deciding how best to evaluate a metaphysical claim. Moreover, this criterion has a degree of objectivity lacking in many other cases. In the first place, it is not up to metaphysicians to decide what lines of thought are initially placed on the table. This is an objective matter based on states of affairs in the first-order disciplines. And while it might be difficult to individuate lines of thought, making it difficult to count how many beliefs have been preserved by a given metaphysical claim, securing agreement on which of two competing metaphysical claims is the more revisionary is usually more straightforward than the adjudication regarding the other second-order criteria.

Some further objections to the aporetic method have been aired within the circle of Aristotelian scholarship. It has been objected, for example, that the best the method can produce is coherence amongst our beliefs. But coherence is not the same as truth; for it is logically possible, indeed it is a frequent occurrence, that a set of coherent beliefs contains elements that are false. So, goes the objection, the aporetic method cannot overcome scepticism.[25] The reply must be that this charge is largely true but beside the point. If one accepts that it is no part of the metaphysician's task to satisfy the sceptic, but rather to resolve aporia in the most truth preserving fashion possible, then this charge can be accepted with equanimity. For the method does provide a means of adjudicating between competing metaphysical claims, and that is all that can reasonably be expected. To ask for more than this – in particular, to ask for deductive proofs from self-evident premises – is to judge metaphysics by the standards of mathematics. But if one maintains, as Aristotle rightly did, that metaphysics is a distinct discipline with distinct problems, methods and standards of evaluation, then its claims must be evaluated by its own lights, not those of a different discipline.

It is also frequently objected that the method as it appears in Aristotle itself relies on 'intuitions'. But, goes the objection, there is no philosophically acceptable reason to give credence to intuitions. In a closely related complaint it is argued that this undue respect for intuitions generates an excessive conservativism which many deem inimical to the philosophical spirit. But while Aristotle might have been vulnerable to these objections, the method as reconstructed here is not. The point of departure of this version of the method are tensions in well-established lines of thought from the various truth-directed disciplines, including the special sciences. These lines of thought are not simply gut-level intuitions or well-entrenched speech habits. Of course there will inevitably be appeal to what one finds plausible at some stage of the process, but the method does not take these intuitions as points of departure as is implied by the objection. And while the method is no doubt conservative in some important respects, it in no way encourages complacency as it enjoins upon the philosopher the task of actively seeking out aporia, an activity which by its very nature brings to light the shortcomings of accepted lines of thought.

A final concluding remark: I have addressed the allegation that answers to metaphysical questions are best provided by the sciences. I have also addressed worries regarding our ability to discriminate between competing ontological claims. What I have not addressed directly is the anti-realist assertion that metaphysical claims lack truth values. The reason for this omission is the belief that the attraction of anti-realism in ontology stems from concerns regarding the epistemology of metaphysics, and I believe I have gone some way to addressing these concerns, thereby removing the strongest temptation to anti-realism. But many will still feel that there are metaphysical questions discussed in the current literature which simply cry out for anti-realist interpretation. I think these sentiments are often well-founded. While I have been at pains to argue that metaphysics as traditionally conceived is necessary and possible, there is no doubt that much current metaphysical discussion is at best unhelpful. The reason for this, I submit, is that much metaphysical discussion is divorced from the task of aporia resolution. If there is a moral to this chapter it is that only those metaphysical claims that offer a solution to an aporia are worthy of serious consideration. Metaphysical claims with no connection to an aporia should be ignored because it is unlikely that one will be able to get any purchase on them, and because they are serving no useful purpose. Here, at least, is common ground with the ontological anti-realist.

3
Evolutionary Biology Meets Scholastic Metaphysics

Introduction

In the last chapter I presented a case for the viability of metaphysics as traditionally understood. Crucial to that defence was an account of the aporetic method. That method enjoins upon us a quite definite procedure. Three points in particular are worth recalling here. First, one should begin one's metaphysical endeavours by identifying aporia arising out of reflection on first-order disciplines. Second, a metaphysical thesis or set of theses is warranted to the extent that it offers the most truth preserving resolution of an aporia. Third, a metaphysical thesis or set of theses can receive further support if it is able to provide truth preserving resolutions of additional aporia unrelated to the first. This approach to metaphysics will be adopted in Part II of this book. Evolutionary biology will be the source of aporia to be investigated.

However, it is not my intention in Part II to simply employ this methodology. It is also my intention to *test* a particular set of metaphysical principles, viz., what one might call the 'first-order' metaphysical theses that received the general assent of the great Scholastic metaphysicians. While there are significant disagreements to be found amongst the Scholastics on precisely how these principles were to be developed and refined, these disagreements were framed against a shared background of broadly Aristotelian commitments. It is this shared framework that is the focus of attention here. And my test is simple: If the basic principles of this framework provide truth preserving resolutions of a range of aporia derived from reflection on biology, then to that extent these principles are rationally warranted, and can be recommended for our general assent. If they do not, then the perennial philosophy will have been found wanting, and alternative metaphysical principles will be needed.[1]

To proceed we need to fix ideas: First, we need an account of the central commitments of evolutionary biology, for it is these which generate the aporia which will be the focus of our attention; second, a budget of aporia specifically drawing on elements of the principles of evolutionary biology; third, a statement of the Scholastic metaphysical principles which will be subjected to aporetic testing in Part II. As a prelude to the third task I sketch an additional aporia, the so-called problem of universals, in order to show the rationale for the adoption of a central element of the Scholastic metaphysical framework, a rationale which retains its force even in a biological context.

The principles of evolutionary biology

An adequate understanding of any theory requires familiarity with the problems it is meant to address. Now evolutionary biologists are particularly concerned to provide an understanding of biological diversity and organismal design. A word on each of these features of the living world is in order.

The diversity and disparity of the living world. Biologists are impressed by the fact that there are so many different kinds of organisms built on such different body plans. The evolutionary biologist seeks to provide some way of making sense of this bewildering variety by finding order in the diversity. But the biologist is also impressed by the fact that this variety is limited. There are many logically possible organisms the biologist can conceive of in 'design space' which she does *not* find in the real world. In fact it would appear that most logically possible organisms never become actual. Thus the biologist also wants to explain why the living world has the pattern it actually has, and why it is not more varied than it actually is. Why, for instance, are there no flying pigs or frogs, or grass eating snakes (there are vegetarian lizards, so why no vegetarians snakes)? If there are eusocial insects, why are there no eusocial birds? Why are there no species with three or more sexes? Why do organisms come in discreet packages – species – rather than all organisms looking the same, or each individual appearing radically different? Why, indeed, has the living world not produced any radically new body plans since the Cambrian 500mya?

Adaptation. A feature of the living world noted by all is the fact that organisms are usually, and often conspicuously, well equipped to deal with their environment. How does this come about? An interesting wrinkle here, however, is that biologists often want to explain why organisms frequently display less than optimal adjustment to their environment. The human eye, for example, though historically used as an instance of intelligent design by a creator, is in fact rather poorly designed from an engineering point of view (retina is at the back rather than the front). Why should this be?

These are the agenda setting questions facing evolutionary biologists, and the theory of evolution is designed to address precisely these issues. With these questions in mind we can turn to the distinctive claims of evolutionary theory. These are as follows:

(1) Evolutionary change has occurred. The living world is not stable, with species coming into and passing out of existence.
(2) All life on this planet descends from a single remote ancestor (i.e., there was no separate or special creation of each individual species) and life has a branching pattern.
(3) New species form when a population splits into two or more groups and these begin to adapt to different circumstances. (Usually a sub-population on the periphery becomes geographically, and so reproductively, isolated from the main population, and begins to adapt to their new and different circumstances.)
(4) Evolutionary change is gradual, not rapid. Offspring that differ radically from their parents due to significant mutation rarely if ever survive to reproduce. All change must be relatively conservative, and so significant changes to a lineage require many small steps taking many generations.
(5) The mechanism of adaptive change is natural selection.

This set of claims has been called the 'received' view, but there is debate about a number of these.[2] Most biologists accept (1) the fact of evolution, and (2) the branching pattern of evolution stemming from a single source (although the shape of life might more closely resemble a mosaic than a tree in single celled organisms). This is virtually universal. Point (3), the theory of speciation (Mayr's contribution), is highly regarded, but not as solid as (1) and (2). It is probably one way new species emerge, but it might not be the only way, or even the most prevalent way. Point (4), the commitment to gradualism, is perhaps more firmly established than it once was now that the excitement that first surrounded Gould's theory of punctuated equilibrium has died down, but developments in evolutionary developmental biology have put this issue back on the table. Point (5) is the element of the received view that has attracted most attention. It is subject to much debate, but most biologists agree that natural selection has at least some role to play in dealing with the explananda outlined earlier. At issue is whether it is the only significant force driving adaptive change, or whether it needs to be supplemented by other forces which might well be more powerful, and whether it can account for the general shape of life. But for the time being we can say this: The *fact* of evolution is established easily. This follows from three readily made direct observations of the living world:

(1) *Phenotypic variety* (Organisms are not identical, but differ within a specific range on a variety of features).
(2) *Differential reproduction* (Organisms do not reproduce in equal numbers. Some produce many more offspring than their conspecifics, many much less, some not at all).
(3) *Principle of heredity* (Offspring resemble their parents more than they resemble other conspecifics).

These facts guarantee that the traits found within a population will change from generation to generation. But if this change is to be adaptive, and if adaptive change is to play a role in speciation, then additional conditions must be met. The change needs to be cumulative, i.e., the same reproductive pattern must take place over many generations. Cumulative selection requires:

(4) Stability in the direction of selection (the same sorts of features need to be favoured over a long period of time)
(5) Each step on the adaptive path must be better than the last (in contrast to philosophical discussions there can be no retreat the better to advance in evolutionary processes)
(6) The right ratio of mutation rate or available variation to selective pressure. If the selective pressure is too hard it will drive

the variation rate down to nothing very quickly, eliminating the chance of further evolution (the experience of animal breeders); but if the selective pressure is too low, then it will not eliminate enough of the variations to make any significant difference to the gene pool as all will survive in equal measure.

Let this suffice as an account of the central claims of the received view of evolution. These will be added to in due course, but we need something relatively definite to give a sense of the theory we are taking as our point of departure.

A budget of aporia

Following Aristotle's lead in the *Metaphysics*, Book III, we need to set out informally the aporia which will serve to guide our metaphysical reflections. A formal statement of these aporia can wait until they are treated in detail. This list is not meant to be exhaustive; but each aporia is meant to highlight a salient metaphysical commitment of evolutionary biology, and each is meant to focus on an issue that does not obviously speak in favour of Scholastic metaphysical principles. That is, there can be no hiding from those grounds which have led many to believe that evolutionary biology is in fact incompatible with Aristotelian metaphysics. The aporia I have in mind are as follows:

(1) The problem of the biological individual

As is clear even from the preliminary account of the received view of evolutionary theory, biology employs the notion of biological individuality at several key stages in the elaboration of its theoretical approach to the living world. In most cases this is unproblematic, for identifying individual biological entities often is perfectly straightforward. However, there are significant puzzle cases that reveal that evolutionary biology does not have as sure a grasp on the notion as is required. As key concepts in the machinery of evolutionary theory presuppose such a grasp, the theory itself must be deemed unsatisfactory. Many attempts have been made to provide an analysis of the concept of the biological individual, but none has proved problem free.

(2) The problem of change

Evolutionary biology is committed to the view that the living world is not stable. It is not just that individual biological entities undergo changes throughout their careers, and that different kinds of individuals have appeared on the scene in the great evolutionary transitions, it is also the case that the Tree of Life itself has changed as species come into and pass out of existence in speciation and extinction events. Moreover, evolutionary biologists assume that biological individuals are not social constructs, but mind-independent entities whose nature is something to be discovered by biological research. It is also commonly assumed that biological entities are fully real in the sense that they are not mere aggregates of physical or chemical components, or fully explicable in terms of the activities of their physical or chemical components. But metaphysicians have long been troubled by the idea that real entities can persist through change. Indeed many have denied that change is in fact possible in real things. The one tradition in metaphysics which does embrace the reality of change, viz., the Aristotelian tradition, is committed to essentialism, thought by many to be incompatible with evolutionary theory. At issue then is whether evolutionary biology is in fact committed to a metaphysically incoherent view of the living world.

(3) The problem of explanation

Evolutionary biology is often lauded for its impressive explanatory power. Indeed, it is this explanatory power which is perhaps the strongest evidence in its favour as a scientific theory. However, explanations in evolutionary biology are essentially contrastive, that is, evolutionary biologists seek to explain why some X has feature Y *rather than some other feature*. To provide such explanations the biologist has to be able to identify what the real alternatives to Y for X were in order to identify the causes responsible for X having Y. But there is no generally accepted view in evolutionary biology circles as to how one identifies the unrealised biological alternatives. But until one has a principled means of doing this, one has no principled way of identifying the cause(s) responsible for X having feature Y, and so evolutionary biology cannot provide the explanations we thought it did.

(4) The problem of value

What is the relationship between biology and ethics? An ancient view, both religious and philosophical, is that ethical values are grounded in human nature. This view is gaining currency again in certain circles, and many have sought to employ evolutionary biology to give this approach scientific respectability. However, science takes itself to be 'value free', and many argue that scientific facts can have no bearing whatsoever on the moral facts (if such there be). Moreover, some take evolutionary biology in particular to show that there is no such thing as human nature, as evolutionary biology is incompatible with essentialism. But, as will be shown in the course of our discussion of the second aporia, evolutionary biology does appear to be committed to essentialism. Disentangling the contradictory claims made regarding the relationship between biology and ethics is the challenge of the fourth aporia.

Such are the aporia that will be examined in detail in Part II. As will become increasingly clear as we proceed, these puzzles arise from the evolutionary biologist's commitment to a set of implicit assumptions, namely, that biological individuals are (1) irreducible, mind-independent entities (a pre-condition of the autonomy of biology), (2) that are subject to change, while (3) remaining intelligible in the sense that explanatory knowledge of these biological individuals is deemed possible. We will have occasion to revisit these implicit commitments in due course, for as we shall see there are problems attending each one of these claims when considered in isolation. But it is worth noting at the outset that Aristotelian metaphysical principles, and their Scholastic refinements, were designed precisely to respect generalised versions of all three claims which biologists find entirely natural. Much effort was required to achieve this because, as a group, the three commitments give rise to at least two classic aporia, the problem of change alluded to earlier, and, with the addition of some considerations regarding the nature of truth, the so-called problem of universals. But because these three commitments remain plausible today, as witnessed by biology, the Scholastic framework developed to accommodate them retains its interest, for it would appear to be the only framework that allows one to retain all three within a coherent system. An entire chapter is devoted to the problem of change, and so can be deferred until later; but a brief word about the problem of universals is in order here.

The problem of universals

For all its associations with medieval philosophy, the problem of universals is not a product of distinctly medieval considerations or commitments but emerges whenever one begins to think about the relationship of language and thought to extra-mental reality. The problem arises whenever one finds a commitment to the following plausible but apparently incompatible propositions:

(1) Explanatory knowledge of the natural order, including the biological order, is possible.
(2) Knowledge is expressed in statements containing general terms and not just names of singulars.
(3) A necessary condition of anyone knowing p is that p is true.
(4) Truth is correspondence.
(5) The natural order contains only singulars.

The tension here is obvious. If knowledge is expressed in true statements, and these statements contain general terms while the natural order contains only singulars, then it is not clear how our knowledge claims can be true because there is a failure of correspondence between our thoughts and reality. Clearly this problem is not confined to biology as it applies to any discipline interested in investigating the natural order.

It is worth noticing that many philosophical systems ignore this problem altogether. Others reject (tacitly or explicitly) one or more of these propositions. So, for example, one might deny (1), claim that knowledge of the natural order is not possible, and embrace either a general form of scepticism or some form of anti-realism in the philosophy of science which maintains, say, that scientific theories are only in the business of allowing for prediction and control of phenomena, or the mathematical representation of a set of experimental laws. But while this approach to scientific theories is not obviously incompatible with some branches of physics and chemistry, it does not sit well with the self-image of biologists.

Biologists tend to be realists about the Tree of Life, and are usually interested in developing the most likely historical explanations for the advent of features on the Tree of Life, prediction and control of future phenomena not usually being at the forefront of anyone's mind.[3] So one might deny (2), and try to show that all knowledge claims can be paraphrased in such a way as to eliminate all general terms. As is clear from the summary statement of the principles of evolutionary biology, it is replete with general terms; but to date no one has been able to show how general terms can successfully be reduced to anything essentially singular, say, bundles of sense data. Indeed this effort at reduction is no longer an ongoing research programme even in radically empiricist circles. So one might care to deny (3), and claim that truth is not a necessary condition of a proposition counting as a genuine piece of knowledge. But this view is not even popular amongst philosophers let alone scientists. As long as one distinguishes between what passes for knowledge and the genuine article, it would seem that one cannot really remove truth from the concept of knowledge. To do so is tantamount to adopting the first option, that is, rejecting (1) and adopting some form of scepticism. So one could deny (4), abandon the 'correspondence' truism about truth, and adopt, say, a coherentist, pragmatic or deflationary theory of truth. Now while biologists are likely to find coherentist theories of *justification* very plausible, biologists are not naturally given to conflating justification with truth. Nor are pragmatic grounds for adopting a theory usually seen as reasons for believing the theory to be true. It is far more natural for a biologist to maintain that their theories are true if they manage to 'get it right' about the particular aspect of the living world they are discussing. After all, this is meant to be the point of testing theories against the available empirical observations. Finally, one could deny (5), embrace a form of Platonism in metaphysics, and claim that the natural order contains not only singulars but abstract objects or universals.[4] Now while all of these options have been explored extensively by philosophers, and while each option has had its defenders, none is particularly inviting from a biological point of view.

Now the interest of Aristotelian Scholasticism is that it sides with the biologist here. The Scholastics insist on retaining all five propositions of this aporia, and so they accept the burden of resolving the tensions between them. The standard Scholastic solution to the problem of universals requires positing (1) *immanent* substantial forms in things themselves (claiming that all singulars, while remaining singulars, come in repeatable types due to the fact that one of their metaphysical parts is a form[5]) and (2) an account of how human beings are able to generate, or otherwise acquire, general concepts that are adequate to entities that are metaphysically singular. After much debate the Scholastics converged on the view that universals in the strict sense are *concepts* with a foundation in reality.[6] Each concept is itself a singular entity (an accident of a particular mind) but in virtue of being predicable of many it is called universal. Concepts, the only entities that are universals in the strict sense, are generated via abstraction from our experience of singulars. These singulars, while remaining singulars, have a tendency to induce our minds to generate concepts applicable to many singulars in virtue of the real resemblances to be found among them.[7] These concepts are arrived at by a process of selectively ignoring the individuating features of singulars while focusing on their shared characteristics. In the realm of concepts the innumerable details of reality are sacrificed on the altar of simplicity and generality.[8] But while the natural order is always richer than our conceptual representations of it, our concepts can be adequate to real singulars because simplification is not falsification, and because at least some of these noticed resemblances are grounded in objective features of things in themselves, namely, their individual substantial form. In non-Scholastic jargon, this means accepting that singulars can be adequately conceptualised as belonging to repeatable kinds or natures.

Now this minimal statement of moderate realism regarding universals is one that all the principal Scholastic figures, from Aquinas to Ockham to Suarez, can accept.[9] It also allows one to respect propositions (1)–(5), so it should be attractive to those who maintain that some account of the relationship between thought and reality that allows for the possibility of knowledge is required. And this is clearly required for those who think that sound metaphysical practice demands that one take the first-order sciences seriously, for all such disciplines, not just biology, presuppose such an account. So in the absence of an alternative account of the relationship between thought and reality, it would seem that there is pressure to resurrect of the notion of a nature and its correlate, substantial form. But these notions do not exist in isolation; they are crucial components of a network of interrelated concepts from which they cannot be extricated without deformation.

My suggestion is that we consider taking this solution to the problem of universals seriously precisely because it solves an aporia in a way that is congenial to evolutionary biology. This means countenancing the notion of substantial forms and the accompanying conceptual framework. The merits of this suggestion are to be tested by further applications of the aporetic method to issues arising from evolutionary biology. I end this chapter then with a statement of certain key principles of this metaphysical framework. I have chosen to focus on the core claims, as well as those which have a bearing on issues that will arise in the course of our investigation of the biologically based aporias.

Skeletal framework of Scholastic metaphysics

The notion of immanent form and nature is but one element of the basic Aristotelian/Scholastic framework. If one looks upon this notion sympathetically then one is likely to find the act/potency distinction, the form/matter distinction, the essence/accident distinction and the substance/accident distinction, similarly congenial. One advantage of adopting this machinery is that it serves as the basis of an overarching philosophical framework embracing one's metaphysics, epistemology and ethics. At issue in this book is whether this framework can survive aporetic scrutiny.

What follows is a set of principles, some more basic than others, to be tested aporetically. The Scholastics had principles covering all topics in philosophy. I restrict this set to the fundamental metaphysical and methodological principles of the Scholastic synthesis that will figure directly or indirectly in Part II.[10]

(1) Being is subject to a number of divisions. One such division is that between real being and beings of reason. Metaphysics is primarily concerned with the former. The latter includes privations and negations, but also imaginary or fictitious entities like gryphons. It also houses logical notions like genus and species, antecedent and consequent, premise and conclusion.

(2) Real being is divided into the actual and the possible. Potency and actuality constitute a complete and fundamental division of being in the natural order.

(3) Actual, real being is divided into the ten categories, the most important of which are substance, quality, quantity, relation, time, place and action.

(4) The principle of substance: Every complete being is an individual substance and an agent.

(5) Parts, powers, properties, acts, accidents and passions are predicated of the substances. (They are ontologically dependent on substances both for their existence and their identity.)

(6) The principle of hylemorphism: Every material substance is composed of two incomplete substantial principles related to each other as potency to act, as matter to form.

(7) The principle of identity: Every real thing has its own determinate essence in virtue of its form.

(8) Form confers specific nature and all its consequences on substances (i.e., its powers, necessary properties, specific activity).

(9) No potency can exist by itself or apart from its composition with act.

(10) Only a being with passive potency can undergo true change.

(11) No potency can actualise itself.

(12) Potency can be known and defined only through its corresponding act.

(13) Substances (essences, natures) are known in and from their operations and accidents.

(14) *Agere sequitur esse*: Activity follows existence, mode of being determines mode of operation, the mode of operation reveals the mode of being, actions reveal the essence, each thing acts according to its own form, every agent acts according to its form.

(15) Such as a thing is, such is its operation and such is the end to which it tends.

(16) Things are essentially distinct if one of them has one or more types of operations the other totally lacks or if one is the analogical cause of the other.

(17) The principle of efficient causality: Nothing can produce itself. Nothing comes from nothing.

(18) The principle of the uniformity of nature: A necessary or natural cause always produces the same effect, one effect, and always acts with the same intensity and in the same manner.

(19) Every effect when known can lead to some knowledge of the existence, power and nature of its cause.

(20) The principle of non-contradiction.

(21) The principle of excluded middle.

(22) The principle of intelligibility: All that is, is intelligible.

(23) The principle of sufficient reason: Everything has either in itself or in another a sufficient explanation for its possibilities, actualities, origin, existence, etc. This is both metaphysical and methodological principle.

(24) The principle of economy. The famous 'razor' has a number of different formulations: One should limit the number of one's explanatory principles; one should choose the lowest level cause adequate to the job when choosing between possible causes; one should choose the cause that requires the least expenditure of energy adequate to the job; one should not multiply distinctions between species, individuals and parts of individuals without necessity.

(25) Truth requires that the mind conform to things.

(26) Truth is one or self-consistent. A truth, whatever its source, never contradicts another truth. One truth may, however, complete the truth or partial viewpoints of another truth.

(27) The universal (that is, the object of a universal concept) is formally in the mind (via abstraction) fundamentally (materially) in similar finite things.

(28) The inference from the possibility of being or action to the actuality of being or action is not valid, but the reverse inference is.

(29) The good of any creature consists in measure, form and order (the primary goodness of a thing is from its form).

(30) The proximate constitutive norm of the morality of human acts is human nature considered completely, both in itself and it all its essential relationships.

(31) The specific natural intrinsic purpose of a nature cannot be known a priori, but only a posteriori. It is learned by the study of the types of goods of which its powers are capable, the types of goods for which it shows a constant and universal tendency and the types of goods for which it has constant and common

needs. These goods, powers, tendencies and needs are in turn known from the observed characteristic activities of a nature.

There is much more to Scholastic metaphysics than this collection of principles. But these form the core of the framework, and it is these that will be directly or indirectly tested in Part II.

Part II
The Turn to Biology

4

Counting Biological Individuals

Introduction

The business of Part II of this book is to bring the core metaphysical principles of the Scholastics into systematic and sustained contact with evolutionary biology. The leading idea is to test the viability of Scholastic principles by seeing if they provide the resources to cope with problems emerging from the first-order disciplines, the natural and social sciences in particular. If they do, then Scholasticism vindicates itself in the marketplace of ideas. If not, then the perennial philosophy needs further development and perhaps revision. This chapter considers the challenge of providing a theory of individuation for living entities. The thesis is that a plausible theory of individuation for biological entities can be forwarded, which relies heavily on the concepts of action and operation. These concepts are central to the Scholastic account of individuals, but are generally absent from the corresponding discussions in the current philosophical and biological literature. The key claim then is that a Scholastic understanding of what it is to be an individual can be deployed to handle a current puzzle in biology, a puzzle that otherwise defies easy resolution. Because this is ultimately the best kind of argument one can offer in support of metaphysical claims, this excursion into biological matters provides a warrant for the Scholastic account of individuals.

Individuating living entities

Philosophical reflection on the biological sciences gives rise to a plethora of questions of a distinctly metaphysical tinge. Perhaps the most familiar metaphysical question associated with biology concerns the ontological status of species. It is natural for metaphysicians to ask whether biological species are mind-independent realities discovered in the natural order, or whether they are artificial groupings that biologists impose upon the living world for their own purposes. And the metaphysician might also ask questions regarding the modal status of these species and other taxa. Is it a contingent fact that the Tree of Life has the shape it does, or is that shape constrained by physics and chemistry? The metaphysician is also likely to ask if entities and explanations in biology can be reduced to entities and explanations in chemistry and physics. But the focus of attention in this chapter is on an important presupposition of these questions.

The presupposition at issue here is that the working biologist can readily identify the individuals of a given species, and distinguish these individuals from their conspecifics, from their biological parts and from the groups they may join. This presupposition is important because it is only when one can confidently identify biological individuals in this way that one can begin to sensibly ask ontological and modal questions regarding the taxa to which they belong, and to which entities they may or may not be reduced. But as we shall see in a moment, identifying biological individuals is not as straightforward as commonly supposed. And this fact raises thorny issues for the biologist as well as the metaphysician.[1] To be sure, in most contexts the working biologist has no problem individuating biological entities. Identifying individual horses, tuna or birches, for example, is not particularly difficult. Each individual horse presents as a discreet unit, neatly separated from other horses and everything else. But not all cases are so straightforward. Counting siphonophoric colonies, particular species of fungi, certain kinds of trees and certain types of slime molds, present real challenges which show up the limitations of our rather vague, pre-theoretical notion of the biological individual. To handle such cases a more sophisticated account of what it is to be a biological individual is required. The problem is that consensus on precisely this point is conspicuously lacking. This lacuna in evolutionary theory places us in the uncomfortable position of maintaining that one of the paradigmatic success stories of the natural sciences, that theory in virtue of which biological phenomena become intelligible, is unable unambiguously to identify the very objects the theory is meant to render intelligible. Meeting this challenge is the first of our aporia.

Before looking at some problem cases and some of the accounts of biological individuality mooted in the literature it is worth underlining two important points: First, that evolutionary biology does indeed presuppose a notion of the biological individual and the organism in particular; second, that biological individuals do indeed exist in the first place to be individuated and identified. Let us take these in reverse order.

If it has proved difficult to light upon an acceptable account of biological individuality one explanation which will suggest itself to metaphysicians is that the problem might stem from the fact that biological individuals are not fully real. If one were willing to countenance the thesis that *composite* entities such as organisms are merely aggregates of simple entities, and that only simple entities can be considered ontologically respectable, then one might see the failure to reach consensus on an account of biological individuality as entirely to be expected. Unger's (2009) sorties paradox of composition as applied to biological entities provides a route to precisely this conclusion. Adapting his case to the present context one can present Unger as arguing that the following three propositions are inconsistent: (1) Organisms exist; (2) Organisms consist of a finite number of cells; (3) If organisms exist, then the net removal of one cell, or only a few, will not mean the difference as to whether the organism exists. Unger suggests that all three propositions are irresistible. But as they are incompatible at least one of them must be false. Unger suggests that the most vulnerable proposition is (1), and so he concludes that composite entities like organisms do not exist, even at the expense of saying that he must now deny his own existence.

I applaud Unger's use of aporetic argumentation, but his recommendation must be resisted by the metaphysician whose method also enjoins a respect for the first-order disciplines. It is not biologically credible to maintain that organisms are mere aggregates lacking full-blooded reality, for this would be tantamount to saying that biology lacks a proper object, and is not an autonomous science. No less an authority on things biological than Ernst Mayr himself goes to considerable lengths to establish that biological organisms are ontologically irreducible, identifying eight characteristics of living organisms that have no parallel in the inanimate world (1982, pp. 36–59). And even those who expect that such an ontological reduction will be effected eventually usually acknowledge that such a reduction would require significant changes to our understanding of physics and chemistry, and most likely include a commitment to downward causation.[2] So our first hope *qua* metaphysicians must be that some account of biological individuality will be forthcoming. Any metaphysical position which leads to the denial of the ontological respectability of biological organisms must be adopted only as a last resort.

As to whether evolutionary biology presupposes a notion of the biological individual, this can be established in the affirmative from the following sorts of considerations:

- The individual organism is commonly taken to be the basic unit of selection.[3]
- Population biology counts births and deaths of individual organisms.[4]
- The concept of 'fitness' applies primarily to individual organisms.[5]
- Adaptations are primarily features of individual organisms.[6]
- The comparative method, a mainstay of evolutionary biology, presupposes that one is comparing like with like. That is, comparative biology must avoid comparing parts with whole organisms, and organisms with colonies if valid comparisons are to be drawn.[7]
- The major transitions in evolution mark the arrival of new kinds of biological individuals, ones which incorporate older biological individuals as mere parts into a higher level whole (e.g., simple replicating molecules, chromosomes, prokaryotes, eukaryotes and multicellular organisms).

These sorts of considerations are enough to show that working with the theory presupposes a secure grasp of the concept of biological individuality. Now consider the problem cases mentioned earlier: How does one count colonial siphonophores like *Nanomia cara*? These creatures look like normal jellyfish but their development is peculiar to say the least. Rather than beginning their careers as a single-celled hydrozoan which develops in various stages into a multicellular body with cellular differentiation, the *Nanomia cara* zygote divides and forms a larva which buds off zooids in a process called astogeny. These zooids remain attached to each other and form a colony, budding off from one of two growth zones in the nectophore region. The colony has zooids of different varieties which together resemble the parts of the true jellyfish. The colony can behave like a single organism, the actions of the individual zooids being coordinated. But there is no process of cellular differentiation as one would find in the normal jellyfish, and the two nectophores have independent nervous systems. So is the colony a single individual, or is each zooid an individual, or each nectophore?[8]

Consider also the perhaps even stranger case of the dictyostelid slime molds (*Myxaomeba*). Their life cycle includes a stage where a fruiting body, a structure with a stalk and a spore container, releases multiple, independently mobile, unicellular spores. At first these spores pursue independent careers, dividing and feeding on bacteria in their immediate environment. Once the supply of bacteria has been exhausted the individual spores begin to starve. At this point they begin to aggregate, first loosely (remaining individually viable) and then tightly. At the advanced stage of aggregation the individual spores take on specific tasks within the now multicellular 'slug', and they lose their individual viability. At this stage the slug presents as a unified organism rather than an aggregate or colony. It eventually transforms into a fruiting body containing spores, and the cycle repeats itself. So how does one count slime molds? Does one count spores, aggregates or fruiting bodies? Or do we have a case here of one individual (the fruiting body) giving birth to many distinct individuals (spores), which eventually unite to form a new single complex organism?

A final problem case worth considering is that of *Armillaria bulbosa*. A clonal population of this fungus covers 15 hectares in a forest in Michigan. While there is no question of its being one continuous body, some mycologists argue it is the largest individual organism on Earth, insisting this fungus is a scattered individual given that all parts of it have been found to be genetically identical. But not all their colleagues are convinced. So how many *Armillaria bulbosa* are in this forest in Michigan, one or hundreds of thousands?

Problematic cases like these illustrate that the notion of biological individuality is not as straightforward as commonly supposed. It is thus not surprising that the literature is replete with discussions of different criteria for biological individuality. The following is a representative sample. It has been suggested that:

(1) Any biological entity which reproduces counts as a biological individual.
(2) A biological individual is the cyclically repeating segment of a lineage.
(3) A biological individual is a biological entity with a distinct genome.
(4) A biological individual is any biological entity with a germ/soma separation.
(5) A biological individual is a living thing with spatial boundaries and contiguity.
(6) If x and y are 'histocompatible' then they are the same biological individual (the immune system fixes individuality).
(7) The sexually fertilized zygote is the biological individual.
(8) The entire mitotic product of the 'bottleneck' stage of the life cycle is a biological individual.
(9) The biological individual is that unit of life within which there is cooperation and no conflict.
(10) Whatever bears an adaptation is a biological individual.

This list of criteria provides a good indication of the sorts of considerations that are at the forefront in the minds of biologists and philosophers of biology when grappling with the concept of the biological individual. But the point for present purposes is that their extensions are not equivalent, and each has its more or less obvious counterexamples. Mules are a problem for (1); single-celled organisms present problems for (2), (4), (7) and (8); clones are the obvious counterexample to (3); groves of aspen challenge (5); the histocompatibility of mothers and foetuses is a counterexample of (6); social insect colonies, as well as meiotic drive and other forms of genomic conflict in multicellular organisms present a challenge to (9); and (10) is in danger of circularity, as the notion of an adaptation presupposes that of biological individuality.

Considerable space and ink could be devoted to a thorough examination of the strengths and weaknesses of these criteria. I do not propose to attempt such an examination here.[9] The justification for this omission is that a noticeable feature of criteria (1)–(10) is that they all isolate strictly *biological* factors in their analysis of biological individuality without pausing to consider what it is to be an individual per se. But it seems reasonable to maintain that providing a plausible analysis of the *metaphysical* notion of the individual per se would be a useful, and perhaps necessary, first step in the provision of a plausible analysis of the biological individual, for the latter is but a special case of the former. So it is to the metaphysical notion of the individual per se that our attention turns. Now while there has been much discussion within the analytic tradition of various issues falling under the banner of 'individuation', little of it is particularly serviceable here. The Scholastics, on the other hand, do provide some useful guidance on the metaphysical nature of the individual per se. This is just the first of many instances of the Scholastic tradition proving its worth.

The metaphysics of individuation

Before examining the Scholastic approach to individuation a few preliminary points need to be addressed. Twentieth-century analytic philosophy contains much discussion of 'individuation'. But the term is used to refer to a number of distinct though related issues that one would do well to keep separate. For example, 'individuation' is often taken to refer to something that human beings do when we 'single out' or 'uniquely identify' some item as a distinct item in thought or reference. In this sense 'individuation' is primarily a topic in the philosophy of mind and language, the challenge being to provide an account of precisely how individuating reference is secured.[10] This is not the sense of 'individuation' of present concern, for we are not concerned at the moment to understand how biologists manage to refer to particular organisms. The issue rather concerns a precondition of this activity of 'individuation', namely that there are items in the natural order to be singled out in the first place. Biologists have no practical problem securing reference to or singling out siphonophores, say, in the sense that in most contexts two biologists can be sure they are talking about the same items in the natural order when they use the term 'siphonophore'.

If this sense of individuation can be styled the epistemological problem of individuation, there is a second set of senses that are metaphysical in import. For example, one often finds discussion amongst analytic philosophers of 'the principle of individuation' where the concern is to state a *criterion of identity* for items of a particular sort. Here the issue has been to determine under what conditions one can say of some item x that it is the same K as y. But again, this is not the sense of individuation of concern to us. Similar remarks apply to discussions of individuation that seek to identify a *principle of persistence* that would allow one to decide whether x remains an instance of K given a change in x of one or more properties over a period of time. This issue will occupy our attention in the next chapter, but it is not the sense of individuation of present concern. Finally, the problem of individuation is often taken to be the attempt to understand what it is that makes an item in the natural order one instance of its kind, distinct from all other items, and the very item that it is.[11] Here one is asking for the *individual essence* of x. And the standard answers, usually formulated with substances in mind, are as follows:

(1) Substances are individuated by their intrinsic properties conceived as universals.
(2) Substances are individuated by their intrinsic properties conceived as tropes.
(3) Substances are individuated by their constituent matter.
(4) Substances are individuated by a combination of their matter and form.
(5) Substances are individuated by their spatio-temporal circumstances.
(6) Substances are self-individuating.

But again, this question is not what the biologist is asking in our current context. Here the biologist is asking not for the individual essence of a siphonophore, say; the issue is whether what is pre-theoretically taken to be a siphonophore should be considered an individual organism or a colony. More generally, at issue is how to draw the distinction between parts of organisms, individual organisms and the groups which individual organisms may join. That is, biologists are not yet clear on what it is to be an individual per se, an issue left unaddressed by contemporary discussions. And it is on precisely this question that the Scholastics prove helpful.

In *Metaphysical Disputation V, On Individual Unity and Its Principle*, Suarez summarises and builds upon over 300 years of sustained philosophical effort to come to grips with questions regarding individuation. Some of these questions map quite nicely on to questions contemporaries are now asking anew. But this is by no means always the case. Suarez distinguishes four main questions regarding the metaphysics of individuals. The first (*Metaphysical Disputation V*, section I, paragraphs 2, 3 and 7) is the constitutive question: What is it for anything to be an individual? The second (all of section I) concerns the extension of the concept: Are there individuals in all the ontological categories? The third (section II) concerns the ontological status of the individuality of individuals: Given that all items are instances of a kind, what does an individual x add to the common nature K to make x this very individual? Finally (section III) there is the causal question: What brings it about that there are individual instances of kinds?

I will return to the constitutive question momentarily, for it is crucial to what follows. But two remarks about the second question are in order immediately. In contemporary discussions of individuation there is a tendency to use the terms 'object', 'individual', 'particular' and 'singular' as little more than stylistic variations of each other. In most contexts this terminological looseness does no real harm. But in the current context it is necessary to distinguish the extension of at least one important and common sense of 'object' from that of the others. While the extensions of 'individual', 'particular' and 'singular' are identical, there are different contrasts these extensionally equivalent terms draw attention to. The contrast with 'individual' is 'group' or 'collection'; the contrast with 'particular' is 'universal'; and the contrast with 'singular' is 'plurality'. Now the term 'object' is usually used to refer to items in one ontological category, namely the category of substance, and so the contrast with 'object' is usually something like 'accident' or 'property'. But while the terms 'object' and 'individual' are often used interchangeably, following Suarez we should not assume that all individuals are objects, as we ought not prematurely to foreclose on the possibility that there are individuals in ontological categories other than substance. That is, I leave it open for the moment as to whether substances are just one species of individual.[12] A consequence of this is that one cannot assume that the defining characteristic of items in the category of substance, namely independent existence, is the defining characteristic of individuality per se. This point is important in our context because a conflation of the notions of 'individual' and 'object' is tempting because the individuals we are concerned with are all substances.

A further comment on the question regarding the extension of the concept of individual unity is in order here. Because it is an open question for the Scholastics as to whether individuals are to be found in different ontological categories, so too it is an open question as to whether there will be a single, univocal concept of individual unity. If there are individuals in various ontological categories then it is highly likely that individuality in one category will not be exactly like individuality in another. This means it is possible that there will be no single set of necessary and sufficient conditions of individual unity. This result should not be met with the standard Quinean response that immediately impugns the usefulness of the concept. If we are to take the Scholastics seriously we must make house room for the notion of analogous concepts, and countenance the possibility that 'individual unity' is one such concept. What is required is that a 'focal' sense of 'individual unity' be identified that will allow one to see principled extensions of this sense to other occasions. A further preliminary remark about Suarez's questions is required. It is to discussions regarding Suarez's rarely distinguished third and fourth questions that many are likely to turn in the first instance when looking for guidance regarding the problematic biological cases identified in the previous section. But this would be a mistake for two reasons. First, this would be to conflate the *cause* of individuation with *what it is to be* an individual, and the answer to the causal question need not be the answer to the constitutive question. And while we do want eventually to be able to answer the question 'What makes x numerically distinct from other members of its kind?' so that one will be in a position to give a principled story about how to count individual organisms, getting clear on the constitutive question 'What is it to be numerically one?' is a necessary preliminary because we need to know what it is to count as an individual unit in the first place.

The second reason it is a mistake to look to discussions regarding Suarez' third and fourth questions is that they are likely to give unserviceable answers. Consider Aquinas' favourite principle of individuation, designated matter. The reason the world presents us with Plato and Socrates, i.e., two distinct human beings, is that the same substantial form has been united with two distinct parcels of matter, the flesh and bones of Plato on the one hand, the flesh and bones of Socrates on the other. Designated matter is thus meant to explain how the world comes to contain distinct individuals of the same kind. The unwary adoption of this approach to individuation suggests that the way to count individual human beings is to count individual human bodies, and thus the way to count biological individuals in general is to count biological bodies. But, barring consideration of the possibility of scattered individuals for the moment, the designated matter criterion comes very close to criterion (5) above, as bodies are standardly thought to be spatially bounded and contiguous. This approach to individuation delivers plausible results when applied to the more familiar biological kinds, but it was developed without any awareness of the peculiar cases that biologists have since identified. For example, it does not sit well with our intuition that a grove of aspen trees contains many individuals despite the fact that the trees share a common root system. If we wish to continue to maintain that a grove of aspen contains many individual trees, then we cannot accept that to be an individual is to be a discrete lump of designated matter. But because Suarez has taught us to distinguish the constitutive nature of individual unity from the cause of such, this is a result we should be able to accept.

Far more useful in our current context is the standard answer to the constitutive question, particularly when combined with a set of interconnected principles concerning the relationship between being and action. The standard answer to the constitutive question, deployed by Suarez in *Disputation* V, section I, is that the focal sense of 'individual unity' is 'that which cannot be divided into units specifically the same as itself'. This constitutive sense of 'individual unity' allows one to distinguish between individual organisms and the groups that individual organisms may join. The idea is that a group of organisms can itself be divided into groups of organisms, and so does not constitute an individual. An individual human being, by contrast, cannot be divided into units that are themselves human beings. Human hearts, livers and brains, for example, are not themselves human beings; they are merely parts of individual human beings. It is this kind of indivisibility that allows one to say that Peter is an individual (in this case an individual human being), while the folk group *Peter, Paul and Mary* is a collection (in this case a collection of human beings).

Applying this criterion of individual unity to the case of individual organisms and groups gives intuitively plausible results, and something similar results from its application to parts of organisms as well. It is not counterintuitive to talk of individual hearts and livers, say, and the reason for this, presumably, is that a heart cannot be divided into units that are themselves hearts any more than a human being can be divided into units that are themselves human beings. But this creates a complication, for now it would appear that one must accept the existence of significantly different types of individuals (parts of organisms and individual organisms), differences that must be recognised by the biologist. Moreover, one might begin to wonder if matters were resolved too quickly when we distinguished groups from individuals by appealing to Suarez's standard constitutive answer. For could one not say of a given collection of organisms that it too must be considered an individual because this very collection is not itself divisible into units that are themselves 'this very collection?'[13] It would seem then that the standard constitutive answer to the problem of individuation leads one to grant the status of individual to parts, wholes and collections, thereby denying one the ability to draw the distinction biologists want to draw.

The Scholastic answer to this difficulty is to accept that individual unities come in different sorts – there are simply different ways of being an individual. But this does not mean that all individual unities are on a metaphysical par. Some individual unities are ontologically prior to others, and what is required is a principled way of determining the ontological order to be found amongst individual unities. And it is here that another crucial set of Scholastic theses comes to the fore.

The fundamental principle of interest to us now is *agere sequitur esse* (action follows being). As Aquinas was wont to say, as a thing acts, so it is; as a thing is, so it acts. Actions and operations reveal something about the nature of the being that performs the action, for the kinds of actions or operations an entity performs depend on what kind of thing that entity is (*operatio sequitur essentiam,* alternatively *operari sequitur naturam vel formam*). But more importantly for present purposes is the claim that the mode of operation of a nature expresses the mode of being of that nature. In particular, Aquinas insists that everything that exists in its own right, i.e., the individual *ens* per se, has or can have, its own independent activity. This, in short, is what it means to be an individual per se.[14] By contrast, those natures which do not and cannot have their own independent activity, but have activities that must be supported by other natures, are not individuals in their own right but depend ontologically in some fashion or another on a distinct entity. Thus a focus on action serves ontology not merely by aiding in the identification of entities and their various powers, but also in the subsequent effort to understand how entities in the various categories are hierarchically related in terms of ontological dependence.

If one takes these principles seriously one quickly arrives at the following suggestion. When counting individuals in any ontological category one should in the first instance focus on whether a putative individual is divisible into units specifically the same as itself, and in the second instance check this result against an inspection of the putative individual's operations, for there are as many individuals in the broadest sense of the term as there are agents. Then, as individuals come in various forms, one should note the independence or otherwise of the operations, for there are as many subsisting individuals, individuals in the strict sense, as there are independent agents. Now these instructions are perfectly general, as they abstract from the nature of the individuals of any particular category. But if one then applies them to biological entities, the category with which we are primarily concerned at present, one should say that x is a biological individual per se if x is indivisible into units specifically the same as itself, and x carries out *independently* the operations of living things. The contrast here with the criteria of biological individuality listed in the previous section is marked. These fail to mention any constitutive criteria of individuality per se, and rarely mention activities or operations at all (focusing rather on structural features of biological entities) while those which do focus on the activity itself rather than its mode.

As suggested at the outset, the merit of such basic principles, and the specific recommendation regarding the analysis of biological individuality that follows from them, is to be determined by their usefulness in dealing with genuine puzzles arising from the sciences.[15] And by this criterion these principles do rather well. For the recommended account of biological individuality immediately provides a principled way of distinguishing between parts, organisms and colonies or groups that is intuitively plausible when applied to both problematic and unproblematic cases. Individual organelles or individual organs (e.g., mitochondrion, hearts) are not biological individuals per se because the constitutive activities of organelles and organs cannot be carried out independently of the rest of the cell or the body. However, when an individual organism enters into a larger system (e.g., a social group or species) it does not join these larger systems as mitochondria join cells or hearts join bodies, for there are activities the organism continues to carry out without the cooperation of the larger systems. Thus the single-celled bacterium or multicellular horse is a biological individual per se because their activities are of the appropriate type. Now groups and colonies, like individual organs, can also be individuated. Because there are activities particularly associated with these levels of organisation, groups and colonies must be recognised as mind-independent entities; but individual groups are not individuals per se because the actions of groups depend on entities each having activities they can carry out independently of the group. This is what distinguishes a particular group or colony from the organism. Thus the independent biological agent provides the focal sense of the term 'biological individual'. Organelles, organs, groups and colonies are biological individuals only in an extended, analogical sense, as these too have actions associated with them, but actions they can perform only because they receive support from biological individuals per se.

Applying these suggestions to the problem cases we get the following results. The fact that a siphonophoric colony can behave in a unified fashion much like an organism shows only that colonies are real biological entities. For that reason a biology that failed to recognise the existence of colonies would fail to provide a complete account of the living world. But no colony is a biological individual per se if any of its parts can perform the operations associated with living things if separated from the colony. Similar considerations apply to the case of *Armillaria bulbosa*. There are as many individual *Armillaria bulbosa* in the 15 hectare patch of Michigan as there are independently acting units carrying out at least some of the life functions of this type of fungus. The fact that these units are genetically identical is no more relevant than it would be when counting twins or clones.

The case of slime molds is more instructive from a metaphysical point of view. For it would appear that we have to recognise as a biological fact of life that an entity's status *qua* biological individual is not necessarily stable. One can start one's career as an individual per se but end it as a part of a larger organism. For the slime mold spores begin as biological individuals per se (feeding and dividing independently); but later they aggregate to the point that they are no longer operating independently. This suggests that the biologist studying these slime molds will have to consider the adaptations and fitness of individual spores *and* individual slugs in order to provide a complete account of these strange creatures. But perhaps the slime mold spore is just an extreme case of a more familiar instance of such a shift in status *qua* individual that has applied to lineages. It is commonly accepted now that the ancestors of mitochondria (α protobacteria), and chloroplasts (cyanobacteria) were independent photosynthetic organisms. These organisms were eventually incorporated into eukaryotic cells as organelles, losing their independent viability in the process.[16] Indeed such shifts in status *qua* individual are one of the generally recognised principles involved in key evolutionary events.[17]

Conclusion

This excursion into evolutionary biology has been an exercise in aporetic metaphysics. The idea has been to test Scholastic principles, not by the lights of contemporary philosophy, but by their ability to help one cope with problems emerging from the special sciences. I suggest that Scholastic principles regarding the constitutive nature of individual unity, and the close connection between being and action, appear to provide the basis of a plausible theory of individuation for biological entities. To that extent one can say that these principles have been 'vindicated'. Of course I cannot claim to have done anything more than to show that these particular principles remain viable after this encounter with a biological problem. A more comprehensive vindication of these principles will require demonstration that they can provide similarly smooth resolutions to further problems. Another such problem is the business of the next chapter.

5
Evolutionary Biology, Change and Essentialism

Introduction

In the last chapter an attempt was made to show that Scholastic metaphysical principles can be deployed to provide a plausible theory of individuation for living entities, particularly individual organisms. Meeting this challenge in a plausible fashion is crucial because of the pivotal role individual organisms play within evolutionary theory in general. I now move on to consider a particular feature of individual organisms, namely, the fact that organisms undergo a variety of changes. In fact evolutionary theory, as the name itself suggests, is deeply committed to the view that virtually no aspect of the Tree of Life is stable. It is not just that individual organisms undergo changes throughout their careers; different kinds of individuals have appeared on the scene in the great evolutionary transitions. It is also the case that the Tree of Life has changed as species come into and pass out of existence in speciation and extinction events.

That evolutionary theory is committed to change being a feature of the living world is hardly controversial. Indeed, as we saw earlier in our account of the received view of evolutionary theory, this is one of its distinctive claims. But when this frank insistence on the reality of change is combined with the view that biological individuals are not social constructs, but mind-independent, ontologically irreducible entities whose nature is something to be discovered by biological research, a difficulty arises. Ever since Parmenides metaphysicians have been troubled by the idea that real entities can persist through change. Perhaps the easiest way to see the difficulty is as follows: If one assumes that an item a has persisted through a change, then a prior to the change is the same item as a at the end of the process (a at $time_{t1}$ is identical to a at $time_{t2}$). But according to Leibniz's Law, if x is identical to y then any property of x must also be a property of y. But if a has undergone a change then it must have some property after the change that it previously did not have, or lost a property it previously had. In either case a at $time_{t1}$ does not have the same properties as a at $time_{t2}$; so by Leibniz's Law a at $time_{t1}$ cannot be a at $time_{t2}$, and so a has not persisted through the change but has been replaced by something else. Generalise this result and one ends up denying that change is possible.

On the back of considerations such as these many metaphysicians have ultimately denied that change as normally understood is in fact possible in real things. They might, like Parmenides, insist that change is an illusion. Another familiar option is to deny that organisms are in fact fully real. Or, adverting to the notions of perdurance and temporal parts, they might offer a radical reinterpretation of objects and nature of change, and insist that, contrary to common sense, only a part of an object exists at any one moment, rather than maintaining that an object endures by being wholly present at every moment it exists. Setting aside the fact that such theories really amount to the denial that change occurs,[1] and setting aside questions regarding the internal coherence of perdurance theories,[2] those inclined to take evolutionary biology and common sense seriously might think: 'So much the worse for those who deny the reality of organisms and change as a real feature of the living world. We ought to have more confidence in the biological sciences and common sense than the reflections of metaphysicians'.[3] There is much to recommend this line of thought. But things are not so easy. For the one metaphysical theory that quite deliberately makes room for the endurance theory of change, viz., Aristotelian essentialism, is precisely the metaphysical theory traditionally deemed to be incompatible with evolutionary thinking in general. So unless we are to ignore metaphysics completely – a course I do not believe is open to us as philosophers for reasons outlined in Chapter 2 – it would appear that one must either deny that evolutionary biology is really committed to the reality of change, or find some way of reconciling evolutionary biology and essentialism.[4] As the aporetic method insists that the metaphysician should take the first-order disciplines seriously, the first option should be entertained only as a last resort. This means we must accept the second horn of the dilemma. Thus the burden of this chapter is to show that reconciling evolutionary biology and essentialism is not nearly as difficult as some imagine. In fact I will argue that far from being incompatible with essentialism, evolutionary biology in fact *presupposes* Aristotelian essentialism inasmuch as the truth of the former requires the truth of the latter. This claim puts me sharply at odds with orthodox philosophy of biology. But I believe this conflict can be resolved amicably once essentialism is properly understood.

To make good this claim it is necessary to begin with an account of Aristotelian essentialism, and to recall the main claims of evolutionary theory. A further preparatory step is to lay out explicitly the standard incompatibilist arguments and some possible responses already mooted in the literature. I can then proceed to the core of this chapter, the presentation of two arguments in support of the thesis that evolutionary biology *cannot* do without essentialism. After floating a suggestion as to what biological essences might be, I revisit the original set of incompatibilist arguments to show that they are easily brushed aside once one is familiar with the outlines of Aristotelian essentialism and the metaphysical commitments of evolutionary biology. But the broader point that emerges from this effort is that it is the metaphysics of the Scholastics that actually sits most comfortably with the deepest commitments of evolutionary biology and common sense.

I turn then to the characterisation of both theories.

Aristotelian essentialism

An adequate understanding of any theory requires familiarity with the problems it is meant to address. This is certainly true of Aristotelian essentialism (from here on just 'essentialism' unless otherwise specified). It is also important for a proper appreciation of essentialism to compare it to the alternative solutions suggested by other metaphysicians (something rarely done in the philosophy of biology literature).

Aristotle's essentialism, the basic outlines of which are accepted by the Scholastics, is the result of the attempt to provide a metaphysical account of what is implicit in our everyday dealings with the world. In particular the essentialist wants to maintain that:

(i) The world contains, amongst other things, mind-independent middle-sized items like minerals, plants, animals and stars;

(ii) These items are irreducible;

(iii) These items can persist through some changes, but not all; and

(iv) These items are intelligible.

As noted earlier, the problem posed by this set of propositions has been to understand how real items can persist through change. Aristotle's solution, designed to respect i–iv, is to accept the following claims:[5]

(1) The world is primarily constituted by individual substances belonging to discrete natural kinds, each kind having its own essential properties.

(2) F is an essential feature of kind K if and only if F is a feature used to define kind K.

(3) The definition of a kind plays two important roles. First, the definition provides the existence and identity conditions of instances of the kind. These allow one to track an instance of a kind through its career and any changes it might undergo by allowing principled answers to questions of the form 'is *a* the same as *b*?' Second, a definition stating the essence of a kind has an explanatory role in that it is adverted to when explaining why an instance of the kind has the properties and behaviour patterns that it does.

(4) There are biological kinds.

(5) (1)–(4) are grounded in the nature of things independently of our thought or representations of them.

Such a theory allows the essentialist to maintain the target theses at the expense of some qualification of Leibniz's Law (it does not apply unqualifiedly across times). (1) and (5) do justice to the reality of middle-sized items mentioned in (i) and (ii); (2) and the first part of (3) accommodates the claim that these items can persist through some changes but not all by distinguishing between essential and nonessential properties, the loss of the latter being consistent with the continued existence of the items through the change, while the loss of the former marks the passing out of existence of the item in question; (2) and the second part of (3) mark a commitment to the intelligibility of these items mentioned in (iv). (4) simply points out that natural kinds are not restricted to items falling exclusively within the domains of physics and chemistry. Crucial to the position is the distinction between essential and nonessential properties. Only if such a distinction is recognised can an entity undergo a change without passing out of existence altogether: accommodating this common sense view is the primary motivation behind essentialism.

Providing a metaphysics which allows one to uphold i–iv is difficult without recourse to essentialism; indeed every competing metaphysical system abandons one or more of the target theses. For example, in asserting the mind-independent nature of middle-sized items the essentialist is at odds with Kant and all forms of constructivism. The essentialist's commitment to (ii) distinguishes him from Plato (who maintained, at one stage at least, that extra temporal and spatial forms alone are ultimately real); from Democritus and other atomists (who reduce middle-sized, and all composite items to aggregates of atoms, the latter alone being fully real); and from Spinoza (who maintained that there is only *one* ontologically basic item). (iii) distinguishes the essentialist from Heraclitus, modern day phenomenalists and trope theorists (who deny the existence of *persisting* objects of any kind). The essentialist's commitment to (ii) and (iii) together distinguishes him from Parmenides, Plato, Heraclitus, Democritus and modern day perdurance theorists who deny that any change is possible in real entities, and from Spinoza who maintains that all changes are merely phase changes of one underlying substance. Finally the essentialist's commitment to (iv) distinguishes him from Parmenides, Heraclitus, Plato and the skeptics and scientific antirealists who all denied that the world of ordinary sense experience is fully intelligible. We shall enter into some of the details of these points later; but it is worth noting at the outset that the rejection of essentialism comes at a high price to ordinary common sense intuitions and to the commitments of evolutionary biology. If one is inclined to believe that individual horses and cabbages, say, are as real as anything can be; that an individual horse and individual cabbage can undergo some changes while remaining a horse or a cabbage respectively, while other changes bring about their respective ends; and if one believes that we can understand something of horses and cabbages (for example, that we can explain why horses have the standard vertebrate limb and cabbage plants can photosynthesise); then Aristotle's essentialism proves indispensable, for the other major metaphysical systems threaten precisely these sorts of claims.[6] Let this suffice as an account of Aristotle's essentialism, and let us now recall the main claims of evolutionary theory.

Evolutionary biology

An account of evolutionary biology has already been provided in a previous chapter, but for the sake of convenience its main claims are reiterated here. Again we start with the questions evolutionary theory is meant to address: evolutionary biologists are particularly concerned to provide an understanding of *biological diversity* and *organismal design*. Biologists are impressed by the fact that there are so many different kinds of organisms built on such different body plans. The evolutionary biologist seeks some order in the diversity by explaining why the living world has the pattern it actually has, and why it is not more varied than it actually is. The main challenge regarding organismal design is to explain how it comes about that organisms display striking, but often less than optimal, adjustment to their environment.

With these questions in mind we can now remind ourselves of the distinctive claims of evolutionary theory. These are as follows:

(1) Evolutionary change has occurred. The living world is not stable, with species coming into and passing out of existence.
(2) All life on this planet descends from a single remote ancestor.
(3) New species form when a population splits into two or more groups and these begin to adapt to different circumstances.
(4) Evolutionary change is gradual, not rapid. Offspring that differ radically from their parents due to significant mutation rarely if ever survive to reproduce. All change must be relatively conservative, and so significant changes to a lineage require many small steps taking many generations.
(5) The mechanism of adaptive change is natural selection.

As previously stated, this set of claims has been called the 'received' view. But that is not to suggest that there has been no serious challenge to various elements of this story. But for present purposes the main point is that change is taken to be a real feature of the living world, and this element of the story is not in contention. Let this suffice as a reminder of the main claims of the received view of evolution. We can now proceed to the grounds for the claim that the two theories are incompatible.

The incompatibilist case(s)

It might not be immediately obvious from the foregoing accounts precisely why the two theories are thought to be incompatible. Many different reasons have been suggested. It is worth spelling out these different lines of thought explicitly.

It is said that essentialism about biological kinds is *not* consistent with evolutionary biology for the following reasons:

(1) Essentialism about species implies species fixism. But species fixism is inconsistent with the view that species evolve. So essentialism about species is inconsistent with evolutionary theory.[7]
(2) Essentialism about species implies clear, non-bridgeable boundaries between species. But this is inconsistent with Darwinian gradualism on two counts. First, no set of properties, at either the level of the phenotype or genotype, has been identified as jointly necessary and sufficient for membership of any biological species. That is, in the field (as opposed to the philosopher's armchair) what one actually finds is such a degree of variation within any species that no clear boundaries between species are found but rather a merging or blending at the edges of one species into another. Second, this degree of variation is a precondition of one species gradually evolving into another, as is demanded by orthodox Darwinism. Evolution between species with clear boundaries would only be possible if nature proceeded by jumps (saltations). But saltations are impossible according to orthodox Darwinian theory.[8]
(3) Moreover, even if the naturalist were to identify necessary and sufficient conditions for membership in a species this would not be to the point. For if an organism were to differ markedly either phenotypically or genotypically from its parents, it would still be classed as a member of the species to which the parents belong. This is inconsistent with essentialism because the properties the

essentialist is willing to countenance as part of an organism's essence must be intrinsic and not relational.[9]

(4) Essentialism is not simply the view that organisms have an essence. It also maintains that this essence has an explanatory role within biology inasmuch as one can explain at least some of the properties of an organism by adverting to the essence of the species of which it is a member. But no essence *with explanatory power* has been identified by evolutionary biology (or any other branch of biology for that matter). Therefore, essentialism is inconsistent with evolutionary biology inasmuch as one claims while the other denies that there are biologically explanatory essences.[10]

(5) It has been argued that biological essences, were they to be discovered, would have no explanatory role in evolutionary biology.[11] In the population thinking characteristic of evolutionary biology, to determine the effects of evolutionary mechanisms one need only advert to statistical laws about the interactions of the individuals in a population. One needs no knowledge of the particular properties of particular individuals. It is only properties of populations that are truly explanatory. 'Describing a single individual is as theoretically peripheral to a populationist as describing the motion of a single molecule is to the kinetic theory of gases. In this important sense, population thinking involves *ignoring individuals. ...* '[12] But in ignoring individuals, one ignores their essences. So essences are explanatorily irrelevant to evolutionary biology.

(6) It is assumed by essentialism that each and every organism has one and only one essence, the essence of the species of which it is a member. But it has been argued that current evolutionary biology favours species pluralism, i.e., the view that organisms can be grouped into several equally real species depending on the species concept employed.[13] What is more, it is claimed that the resulting species do not coincide. That is, it is not the case that reproductively isolated groups coincide with groups with common ancestors and groups subject to the same environmental selection pressures (groupings arrived at using the biological, phylogenetic and ecological species concepts respectively). As one and the same organism can fall into more than one group, and as no one of these groupings is privileged, it would seem that

an individual organism can have more than one essence, contra essentialism.

But the incompatibility thesis has been contested on the following grounds:

(1) Bernier argues that essentialism is not incompatible with evolutionary biology because species fixism, properly conceived, is not incompatible with one species giving rise to another distinct species via standard evolutionary processes.[14]

(2) D. Walsh, relying on Pellegrin,[15] D. M. Balme[16] and J. Lennox,[17] argues that essentialism is not incompatible with evolutionary biology because essentialism properly conceived does not imply species fixism of any description. He writes: 'On Aristotle's scheme essences or natures are not transcendent fixed "ideas"; they are goal-directed capacities immanent in the structure of the organism'. These natures '...could change over time in just the way we have come to think that species do'.[18]

(3) Walsh argues, contra Sober, that evolutionary biology cannot rely simply on population thinking while ignoring individual organisms and their properties. While evolutionary change can be described as changes in gene frequency in a population (as Sober suggests) one cannot explain why such changes are adaptive without adverting to features of individual organisms, in particular their developmental systems and phenotypic plasticity. As these features are plausibly regarded as the essential nature of organisms, and as explaining adaptations is part of the raison d'être of evolutionary biology, evolutionary biology cannot fulfil its explanatory ambitions without presupposing essentialism. 'Recent evolutionary developmental biology shows that one cannot understand how natural selection operating over a population of genes can lead to increased and diversified adaptation of organisms unless one understands the role of individual natures (essences) in the process of evolution'.[19] Therefore essentialism is not inconsistent with evolutionary biology.

This collection of arguments is not exhaustive, but it includes the most pressing points advanced on both sides of the debate. It is worth noting immediately that the incompatibilist arguments are not consistent. Some deny there are biological essences (1–3); others are willing to countenance essences but deny them explanatory value (4 and 5); still others claim that organisms can have *more than one* biological essence, each possibly having explanatory value in some context or another (6). The same can be said of the arguments on the other side inasmuch as there is a difference of opinion as to whether species fixism is indeed a problem. Some claim that it is not (1), while others, at least by implication, assert that fixism would be a problem if it were entailed by essentialism (2). I take these inconsistencies on both sides of the house to indicate both the complexity of the issues and the need to return to first principles. The first principle shared by both theories is a commitment to the reality of change in the living world. It is on this shared principle that I build two presupposition arguments intended to show that evolutionary biology actually requires the truth of essentialism.

Two presupposition arguments

As stated at the outset, I maintain that both evolutionary biology and Aristotelian essentialism have independently established claims on our allegiance. Consequently, on the assumption that truth is one, it is methodologically appropriate to start with the assumption that the tensions between the two are not genuine but merely *prima facie*. Of course if this thesis cannot withstand scrutiny one will have to accept that the tensions are genuine and a choice between the two will be forced. It is my view that this choice can be avoided.

It is important at the outset to be explicit about my limited aims. I am not concerned here to defend directly either the received view of evolution or essentialism. The question here is only as to their compatibility. I want to know whether the truth of either would imply the falsity of the other. As far as the argument of this chapter is concerned *both* evolutionary theory *and* Aristotelian essentialism *might very well be false*. I happen to think both are true, and that this position will be supported in some measure if I can but show that they are at least compatible.

With this in mind I now present two arguments which invite the conclusion that evolutionary biology presupposes Aristotelian essentialism. Both are based on considerations drawn from reflection on the very problem of change that motivated essentialism in the first place. The gist of these considerations is as follows: If organisms are ontologically irreducible to entities of physics and chemistry; if biological species are natural groups of such organisms; if such species can undergo some changes without passing out of existence; and if one is willing to accept that speciation and extinction events do occur, then essentialism is forced – for an entity can persist through change only if it retains its essential properties while shedding or gaining an accident property. Now it would appear that the only claim in these reflections at which some biologists might baulk is the claim that species are natural groups. For as we saw in the previous chapter, the autonomy of biology as a science requires organisms to be ontologically irreducible to physics and chemistry, and we have yet to see good reasons to deny this claim. Moreover, all agree that a species can, say, increase or decline in numbers, or broaden or decrease its range, i.e., change in some respect, without ceasing to exist. And of course no biologist is going to question the propriety of speciation and extinction events. The two arguments to follow are thus designed to show that evolutionary biology will have great difficulty in discharging its own self-imposed explanatory goals if it abandons the view that species are natural groups.[20]

An argument from diversity:

(1) Evolutionary biology's fundamental claim with respect to biological diversity is that species have diverged to take advantage of the various ecological opportunities afforded to them. Ancestral species have given rise to distinct daughter species by a process of descent with modification, which results in the emergence of new biological forms and the expected degree of biological diversity. In short, biological diversity follows upon speciation events.

(2) Setting aside questions regarding the various possible mechanisms of speciation, it is customary within evolutionary biology to take the following view of the origin of species. Once ancestral species A has cleaved into two new daughter species B and C, ancestral species A no longer exists, and daughter species B and C have come into existence (there have been two speciation

events and one extinction). Moreover, B is not C, and neither is a continuation of A.[21]

(3) This account of the origins of biological diversity presupposes that change is a real feature of the living world. In particular it presupposes that distinct species really do come into and pass out of existence. So speciation and extinction events are not illusory. Nor are they simply a function of our naming conventions – for mind-independent diversity cannot be explained by mind-dependent, i.e., non-natural, entities and processes. Furthermore, the biologist cannot maintain that speciation and extinction events are merely a function of a new arrangement of subatomic particles, or merely a phase change of an underlying substance, or temporal parts of an unchanging Tree of Life without abandoning the ontological irreducibility of organisms or the reality of change.

(4) It is possible to maintain that A, B and C are distinct, natural species only if the existence and identity conditions of each are distinct.

(5) This point is generalisable to cover all speciation and extinction events.

(6) But the existence and identity conditions of x specify the Aristotelian essence of x. So,

(7) Biological species, in virtue of having existence and identity conditions, have an essence.

The upshot of this argument is clear enough: The standard account of biological diversity provided by evolutionary theory presupposes essentialism. Note that this argument is built on the fact that species *do* come into and pass out of existence, a fact often thought to be inimical to essentialism. In fact quite the reverse is the case. Only if species have distinct essences can one say in a principled fashion that one species no longer exists and that two new distinct species have arrived on the scene, and one needs to be able to say this if one is to give the standard account of biological diversity.

An argument from organismal design:

(1) Evolutionary biology's fundamental claim with respect to organismal design is that many features of organisms are adaptations.

(2) An adaptation is a derived character or trait that evolved because it improved relative reproductive performance.

(3) Crucial to present purposes is the contrast between derived and ancestral characters. A trait or character is termed 'ancestral' if it is possessed by an ancestral species shared by related daughter species. A trait or character is termed 'derived' if it evolved after the ancestral trait in the lineage.[22]

(4) To determine whether a trait is derived one needs to know something of the transition from the ancestral to the derived condition of the character. That is, one needs to know the trait's phylogenetic history.

(5) To track the phylogenetic history of a trait the biologist employs phylogenetic trees.

(6) For a phylogenetic tree to be genuinely illuminating it must represent real relationships obtaining between natural species.[23]

(7) The standard relationship represented by a phylogenetic tree is that of an ancestral species A cleaving into two or more daughter species B and C.

(8) And as seen in the argument from diversity, the standard interpretation of this process assumes that after cleavage species A no longer exists, and species B and C have come into existence (there have been two speciation events and one extinction). Moreover, B is not C, and neither is it a continuation of A.

(9) Thus in order to maintain that a trait genuinely is an adaptation the biologist must assume that distinct, natural species really do come into and pass out of existence. That is, speciation and extinction events are not illusory, nor simply a function of our naming conventions – for mind-independent adaptations cannot be explained by mind-dependent, i.e., non-natural, entities and processes. Furthermore, the biologist cannot maintain that speciation and extinction events are merely a function of a new arrangement of subatomic particles, or merely a phase change of an underlying substance, or temporal parts of an unchanging Tree of Life without abandoning the ontological irreducibility of organisms or the reality of change.

(10) It is possible to maintain that A, B and C are distinct, natural species only if the existence and identity conditions of each are distinct.

(11) But the existence and identity conditions of x specify the Aristotelian essence of x.

(12) In order to maintain that a trait is an adaptation the biologist must assume it is a feature of a species with an Aristotelian essence.

Again, the upshot of this argument is clear enough: The standard account of what it is to be an adaptation presupposes essentialism. Phylogenetic trees can be genuinely illuminating only if they represent real relationships between natural groups which come into and pass out of existence. But it is only if species have distinct essences that one can say in a principled fashion that one species no longer exists and that two new distinct species have arrived on the scene. Thus the standard accounts of biological diversity and organismal design both presuppose essentialism.

What are these essences, and are they explanatory?

It would certainly smooth the path of the arguments from diversity and organismal design if some account of these alleged biological essences were forthcoming. My main suggestion regarding biological essences is that they are found not in the genotype or the phenotype but in the species specific developmental programmes that map genotypes onto phenotypes. The key claims in this suggestion are that (i) only a portion of an organism's genome determines its species (not all of it); (ii) that developmental control genes (i.e., genes that control the expression of other genes) determine the developmental pattern of an organism; and (iii) that these developmental patterns are 'lineage specific', i.e., shared by individuals of the same biological species understood as a smallest diagnosable cluster of organisms related by ancestry and descent.[24] On this suggestion two organisms belong to the same species and have the same essence if they share the same developmental programme regardless of how else they might differ. If a population of such organisms maintains the same developmental programme over several generations then no extinction or speciation event has occurred, regardless of any other changes that might have taken place.

Perhaps the most striking thing about this suggestion is that its plausibility is granted even by those who are not usually considered friends of essentialism. John Dupré, for example, has written:

> It might reasonably be asked here whether these epigenetic mechanisms might not themselves serve as essential properties. And I think that if, as I speculated earlier, there are species for which these provide the best account of species coherence, we would have here perhaps the best candidates in biology for real essences.[25]

One reason for taking species specific developmental programmes as serious candidates for biological essences is that they have great explanatory potential, an essential feature of Aristotelian essences. A developmental control gene can be seen as a selector switch that makes choices from a range of potential developmental fates. These switches are responsible for the 'universal' properties of phenotypes. And these switch points allow for phenotypic alternatives that can become subject to selection pressures. Moreover, M. J. West-Eberhard has fixed upon these features of developmental programmes in the elaboration of her developmental plasticity theory of speciation. She writes:

> developmental plasticity in trait expression within a parent population can predispose descendent sister populations to speciation by facilitating the intraspecific evolution of contrasting specializations. The individuals expressing these specializations begin to show breeding separation... This creates two breeding populations, each one with one of the contrasting alternatives *fixed*. Phenotypic fixation... promotes further divergence.[26]

The main lesson she draws from this line of thought is that 'Phylogenetic gaps could have a developmental origin'.[27] R. Raff, for one, would concur:

> Novel features arise in animal evolution as a result of modifications of developmental pattern.[28]
>
> Most of what goes on in the development of a new descendent species will utilise the same standard parts as the parent species.

Novel forms will arise mostly from the modifications of existing modules in development.[29]

Now it is not the business of the metaphysician *qua* metaphysician to defend the empirical adequacy of this thesis regarding biological essences. Of course there are outstanding questions that need to be addressed. Will this approach work for all organisms? Are developmental programmes as invariant as this proposal suggests? These are empirical questions best left to biologists.[30] From a metaphysical point of view it suffices to make the suggestion, and draw attention to the attractions of the view. But *qua* metaphysicians we can say at least two things here. First, if this particular suggestion does not hold up under scrutiny, something else will have to be found to play the role of essences if evolutionary biology is to meet its self-imposed explanatory objectives. Second, while its confirmation lies ultimately in the hands of biologists, it is to be noted that the claim that species specific developmental programmes are biological essences does not fall to any of the original incompatibilist objections rehearsed at the outset of our discussion, and this serves as a kind of corroboration. I conclude, then, with a brief review of those original incompatibilist arguments with this thesis in mind.

Replies to incompatibilist arguments

Some of the replies to the incompatibilist arguments will be clear enough from the foregoing discussion. For example, it has already been pointed out that there is nothing in Aristotelian essentialism that implies species fixism, i.e., that species cannot evolve. In fact essentialism is required to allow for genuine change in the living world. Similarly, we can reject the second objection on the grounds that an organism's species specific developmental programme is that in virtue of which it belongs to a particular species, happy that this allows for the full range of phenotypic variability found in real populations. Until this suggestion is defeated on empirical grounds there is, contra this objection, an empirically plausible candidate for the role of biological essences.

The third objection is curious in that it appears to undercut evolutionary biology itself. For if offspring are always placed in the same species as the parent regardless of genotypic, phenotypic or developmental differences, as the argument alleges, then speciation events would be impossible. This is an argument against the received view of evolutionary biology, not essentialism.

As to the fourth objection, which granted essences house room within biology but denied them explanatory power, perhaps enough has already been said. One of the main attractions of the thesis that species specific developmental programmes are biological essences is precisely their explanatory power, so the objection is simply false.

To the fifth objection – Sober's argument that the properties of individuals can be ignored in population thinking, so the essential properties of individuals (if they existed) are not explanatory – it can be countered that the statistical properties of populations are ontologically dependent upon the properties of the individuals that make up the population. So at some explanatory stage the properties of individuals must be factored in. Their essential properties will be among those adverted to in the course of this level of explanation. And there is no reason to think developmental programmes will not be involved, at least indirectly, in these explanations.

Finally, what are we to make of the claim that one and the same organism can belong to several, equally real biological species, so that one and the same organism can have several, equally real essences? This last objection falls foul of the principle of non-contradiction and so is charged with incoherence. If one and the same organism had more than one essence, then it would have more than one set of existence and identity conditions. But this would allow it to possess under one set of conditions a property which it does not have under another – a violation of the principle of non-contradiction. This result can be avoided in one of two ways: Either one can deny species realism, but at the cost of compromising the explanatory goals of evolutionary biology; or one might claim that two or more organisms can occupy exactly the same space at the same time, a claim few biologists would find intelligible.[31] It is much more plausible to avoid the contradiction altogether and maintain that each and every organism has one and only one developmental programme, and so each and every organism has one and only one essence.

So I conclude that the original incompatibilist objections leave unscathed the suggestion that biological essences are species specific developmental programmes. This in turn makes the acceptance of the two presupposition arguments easier to countenance. But the crucial point upon which all else depends is the commitment to the reality of change shared by evolutionary biology and essentialism. It is this shared metaphysical commitment that binds the evolutionary theorist to the essentialist.

6
Evolutionary Biology, Modality and Explanation

> ...the contemplation in natural science of a wider domain than the actual leads to a far better understanding of the actual. (Eddington, 1928, pp. 266–267)[1]

Introduction

So far in Part II I have been at pains to deal with aporia related to evolutionary biology's commitment to the mind independence and irreducibility of biological individuals, and the claim that such individuals, and much besides, are subject to change. In the last chapter I also attempted to show that evolutionary biology's self-imposed explanatory tasks could not be accomplished without accepting Aristotelian essentialism. The main idea has been to subject Scholastic principles to aporetic testing. What has emerged so far is the deep compatibility between evolutionary biology and the metaphysical principles of Aristotle and the Scholastics. This chapter develops these themes further by focusing on another issue arising from evolutionary biology's claim to provide adequate explanations of both biological diversity and organismal design.

That evolutionary biologists see themselves as in the business of providing explanations for a range of biological phenomena is obvious enough. What is also clear to anyone familiar with contemporary philosophy of science is that the concept of explanation itself has proved deeply problematic. At the moment there is no consensus on a unifying account of the nature of explanation.[2] However, I propose to ignore this particular philosophical difficulty for the moment. In

116

keeping with the aporetic method, I am going to take as my point of departure the standard view in biology that to explain something is to have identified its cause.[3] The causes standardly appealed to are *proximate* causes (i.e., those causes operating within the lifetime of a single organism including mechanical explanations of how a trait or function is realised in the tissues of an organism, and developmental explanations of the construction of organisms) and *ultimate* causes (i.e., those causes of micro and macroevolution operating over many generations including selection accounts of the emergence of a trait, and phylogenetic history). The issue regarding explanation in biology is thus not what constitutes an explanation, but the challenge of identifying the causes in play in any given circumstance. But as we shall see, meeting this challenge is not just an empirical problem, but one which raises deep metaphysical complications. This chapter focuses on the need to delimit and classify nonactual biological forms when offering explanations in evolutionary biology. Such a classification scheme sorts actual and nonactual forms according to their modal status. Such a sorting has been attempted by theoretical morphologists, but these efforts have paid insufficient attention to the metaphysics of modality. Contemporary approaches to the metaphysics and epistemology of modality are also found wanting. The primary burden of the chapter is to suggest yet again that the necessary intellectual resources are to be found in the metaphysical principles of the Scholastics regarding the nature of modality.

Taxonomy and explanation

It is hard to overestimate the importance of taxonomy to our understanding of biological phenomena. One need only appreciate the dependence of the comparative method on a reliable taxonomy, and the centrality of this method to much of biology, to see that an adequate delimitation and classification of biological forms is the foundation on which the more eye-catching branches of biology are built.[4] However, the theory and practice of delimiting and classifying *actual* biological forms is not the focus of attention here. Our present concern is the much less familiar but arguably no less important challenge facing those wishing to sort actual and nonactual biological forms into the various modal categories. Most pressing is the need to light upon a principled means of determining in the case of any given nonactual form if it is nonactual but biologically possible, or nonactual because impossible. And if a given form is impossible, we also need to be able to determine if its impossibility is due to biological, physical, metaphysical or logical factors.

It is not immediately obvious why a sorting of biological forms on modal criteria is required, so the first section of this chapter is devoted to motivating this particular taxonomic project. It is here that the aporetic method developed in Part I of this book comes to the fore. The second section establishes that the intellectual resources necessary to carry out the required sorting are not available in contemporary approaches to modality. The final section argues that the Scholastic approach to modality provides positive guidance on how the required sorting is to be carried out in practice. But again the ultimate message is the same: Scholastic metaphysical principles allow for the most economical resolution of aporia stemming from first-order sciences.

The modal categories and contrastive explanations

The need to sort actual and nonactual biological forms on modal criteria stems from the fact that explanations in evolutionary biology are essentially contrastive. Very often the contrasts of interest are those obtaining between an *actual* form and a range of other suitably chosen *actual* forms. But this is not always the case. Very often a biologist will not fully understand an actual form without first comparing it to a range of nonactual forms. This is particularly so with respect to evolutionary biology's two self-imposed explanatory tasks, viz., explaining biological diversity, and explaining organismal design. In such cases the biologist is always asking why a given trait or feature obtains *rather than some other trait or feature*. Dennett offers a particularly clear expression of this point:

> Any acceptable explanation of the patterns we observe in the Tree of Life must be contrastive: why do we see this actual pattern rather than that one – or no pattern at all? What are the nonactualised alternatives that need to be considered, and how are they organised? To answer such questions, we need to be able to talk about what is possible in addition to what is actual.[5]

Those interested in the metaphysics of causation will note a significant point here. As explanations in biology are causal, it would appear that biologists are lending tacit support to a controversial but plausible claim regarding the number of roles one must posit in any causal relation. In the view adopted here the form of a causal relation is always 'c_1 rather than c_2 or c_3 or c_n causes e_1 rather than e_2, or e_3 or e_n' despite the fact that the contrasting causes and alternative effects are usually left unexpressed.[6]

Now it is the need to specify the modal character of the alternative effects in biological explanations that is the focus of attention here. To make progress on explaining why the Tree or Mosaic of Life has the shape it does, for example, the biologist needs to explain why life on Earth presents precisely *this* degree of diversity and no more. And this question arises precisely because biologists have hitherto assumed, quite plausibly, that life on Earth has *not* exhausted all the genuine biological possibilities. Indeed some go much further and insist that '[t]he actual animals that have ever lived on earth are a tiny subset of the theoretical animals that could exist'.[7] If this is so, then it is perfectly natural to wonder why just these organisms made it to actuality while the vast majority of conceivable organisms did not. As McGhee puts it, when explaining biological diversity, '[t]he ultimate goal is to understand why extant form actually exists and why non-existent form does not'.[8] But addressing this question successfully presupposes the ability to trace the boundaries between distinct modal categories. First, one must be able to distinguish between the nonactual but possible forms and the nonactual because impossible forms; second, among the impossible forms, one must be able to distinguish between the biologically, physically, metaphysically and logically impossible forms, each form of impossibility being a distinct modal category with different implications for explanation. For the explanation of why the world contains ambulatory pigs and carnivorous snakes, say, but *not* flying pigs or vegetarian snakes, depends in large part on whether flying pigs and vegetarian snakes are nonactual but possible organisms or simply impossibilities. The first suggests that nature has simply not got round to making flying pigs or vegetarian snakes – perhaps the necessary mutations in the pig or snake populations have yet to occur, or perhaps incipient flying pigs and vegetarian snakes have arisen but have been actively selected against in past environments. But if such organisms are simply impossibilities, then flying pigs and vegetarian snakes were never an option because such organisms cannot be made. Natural selection would then have nothing to do with such organisms failing to reach actuality. I will return to this explanatory difference shortly, but the main point for present purposes is that for the biologist to fully understand extant forms she must have a reliable grasp of the modal character of the extant forms and that of the alternative effects. Efforts in this direction are found in theoretical morphology, the aim of which is 'an understanding of biological diversity, framed in terms of the boundaries between the possible and the actual and the possible and the impossible'.[9]

Being able to trace the boundary between the nonactual but possible and the nonactual because impossible is also crucial to the other explanatory task of evolutionary biology – accounting for organismal design. The leading idea here has been to advert to the notion of adaption. And the task has been to establish in any given case whether the trait in question is an adaptation, the operating assumption being that the sheer existence of a trait does not imply adaptive significance. To establish that a trait is an adaptation one needs to show (amongst other things) that it was selected from a range of alternative values for the trait, for a trait is an adaptation, not if it is absolutely optimal, but if it is optimal vis-à-vis a range of real alternatives. As it is usually impossible to know empirically what the variations within the population in the ancestral environment actually were (because these may have long been eliminated from current populations) biologists must find some other way of identifying a plausible range of alternative values for the trait. Again, to do this biologists must construct a 'morphospace' representing hypothetical yet potentially existent morphologies. The idea is that they can then compare this set of possibilities with reality to see which of the possible forms are common, rare or nonexistent. They can then carry out a functional analysis of both existent and nonexistent forms to see if the existent forms are of adaptive significance. But again the point for present purposes is that meeting the second of the self-imposed explanatory tasks presupposes that the biologist has a reliable grasp of the modal character of the alternative effects. Without such a grasp any conclusions based on comparisons of functional analyses are very likely to be unsound as one might include within the morphospace conceivable but biologically impossible values for the trait. Such a conflation of the biologically possible and impossible could result in a trait's adaptive significance being missed, say, because a conceivable though biologically impossible value for the trait performs better on a functional analysis than the actual value despite the fact that the actual value is the best of the real alternatives.

So much is clear about the contrastive nature of biological explanations, and why a reliable means of sorting actual and nonactual forms on modal criteria is required. But what is also clear, yet seldom discussed with the requisite urgency, is that as yet there is no widely accepted means of sorting forms into modal categories. Indeed one of the great controversies in contemporary biology – the dispute concerning the principal driver of evolutionary change – turns in large part upon different intuitions regarding what is biologically possible. Orthodox neo-Darwinians, and their bug-bear, Stephen J. Gould, often talk as though what is biologically possible is virtually limitless. This modal claim suits both camps well. It fits nicely with the neo-Darwinian view that natural selection is the main, and perhaps the exclusive, explanatory tool of evolutionary biology, for it is natural selection that allegedly winnows this vast expanse of genuine possibilities down to the few lucky winners that make it to actuality. And it fits Gould's intuitions that life on Earth would be very different if history, with all its contingencies, were to be rerun. But both Gould and the neo-Darwinians face a serious challenge from another school of thought, with roots in the work of D'Arcy Thompson, which maintains that the range of genuine biological possibilities is actually severely limited, and so the explanatory import of natural selection or historical contingencies is greatly diminished. In his provocatively titled *Evolution without Selection* Lima-de-Faria writes:

Biological evolution exists for the simple reason that it could not be avoided. The proton, the neutrino and the boson contained at the dawn of the formation of the universe the properties that would make later plant and animal evolution inevitable. Moreover, but most important, this biological evolution arose as a prisoner of the rules and principles guiding the initial construction of energy and matter and as such cannot follow any lines of development except the narrow ones imposed by this initial restrictions and canalisation. Biological form and biological function are the products of the mould of form and function already present in the quarks and leptons, or in any other of the elementary particles. [10]

In this line of thought physico-chemical constraints are accorded pride of place with natural selection and historical contingencies reduced to a marginal role. But the point for present purposes is that there is no principled way to decide between the neo-Darwinians and Gould on the one hand, and the structuralists on the other, without an adequate grasp of the modal character of both the actual and alternative effects.[11] One illustration taken from Lima-de-Faria will suffice. Why do humans smile? Neo-Darwinians will say that at some stage of hominid evolution there was selective advantage to be had from the ability to smile, and this trait was selected from a range of alternative traits. That is, it was biologically possible for humans *not* to have the ability to smile, but this possibility was selected against. A structuralist like Lima-de-Faria rejects this explanation precisely because he maintains that the ability to smile in humans is *not* one of many alternatives but was forced by physico-chemical constraints having nothing whatsoever to do with selection.

At present there is no principled way of deciding who is right about the modal status of traits like smiling. This state of affairs places evolutionary biology in a particularly unwelcome position. For no putative explanation of a given phenomenon is warranted if it cannot eliminate relevant alternative explanations in a principled fashion. So, as things stand, evolutionary biologists appear to be committed to the following inconsistent set of individually plausible propositions:

(1) One of the great merits of the neo-Darwinian synthesis is its extraordinary explanatory power. Indeed this explanatory power is one of the best reasons we have for accepting this theory as a true account of the living world.

(2) Explanations in evolutionary biology are couched in terms of contrastive causes and alternative effects.

(3) Contrastive explanations presuppose a reliable grasp of the modal character both of the effect and the nonactualised alternatives, but

(4) There is currently no consensus on how to determine the modal character of the effect and the nonactualised alternatives.

When presented in this fashion the challenge facing evolutionary biology is obvious enough. The explanatory power of evolutionary biology must be illusory if (2), (3) and (4) are true. I do not wish to abandon evolutionary biology, nor do I think this is a serious option. But I see no way of plausibly denying (2) and (3). The only way to deal with this challenge is to rectify the state of affairs described by (4).

Of course this challenge is particularly embarrassing for biology. But there are good grounds for claiming that the real shame is philosophy's. For in a plausible view of the division of labour between the disciplines, it falls to metaphysics to delimit the possible, necessary and impossible, while it is the business of the special sciences, in this case biology, to determine empirically which of the possibilities have been actualised in the real world.[12] But metaphysics has not lived up to its part of the bargain. In the next section I will show that none of the approaches to the ontology and epistemology of modality currently popular in mainstream metaphysical circles is even remotely helpful when considered from a biological point of view.[13] Little wonder that biologists themselves have stepped into the breach. Unfortunately the most sophisticated efforts to date in theoretical morphology are inadequate as morphospaces are determined primarily by the dictates of geometry, and are developed without reference to real organisms. This is clear enough from the following extended passage from McGhee:

> ... the goal is to explore the possible range of morphological variability that nature could produce by constructing *n*-dimensional geometric hyperspaces (termed 'theoretical morphospaces'), which can be produced by systematically varying the parameter values of a geometrical model of form. Such a morphospace is produced without any reference to real or existent organic form. Once constructed, the range of existent variability in form may be examined in this hypothetical morphospace, both to quantify the range of existent form and to reveal non-existent organic form. That is, to reveal morphologies that theoretically could exist (and can be produced by the computer) but that have never been produced in the process of organic evolution on planet Earth. [14]

Unless one believes that the constraints of geometry exhaust the constraints placed on the living world, so that the *biologically* possible extends to the *geometrically* possible, current efforts in theoretical morphology cannot be accepted as satisfactory.[15] There is no doubt that such techniques do identify forms that are 'theoretically' possible in some sense; but 'theoretically' here does not draw the needed fine-grain distinctions between the various ways in which a form might be possible.

The poverty of orthodox approaches to modality

For reasons outlined in the previous section we are in the market for principled method of sorting actual and nonactual biological forms according to their modal character. We need to be able to do this in order to tell which of a variety of causes might legitimately be adverted to as an explanatory mechanism for any given biological form. Intuitions amongst biologists on this score differ dramatically, so I take it that a return to first principles is required. It is very difficult to discern what is possible and impossible in any domain if one does not know what features of reality ground the modal facts in that domain. So I take it that we need both an ontology and epistemology of the biological modalities. Unfortunately we do not get much help on these matters when we turn to the contemporary theories of modality developed by mainstream metaphysicians. Space considerations allow only a brief overview of these approaches, but even a brief word on these matters is enough to appreciate that contemporary metaphysics is not going to come to our aid.

Let us remind ourselves of our task: To achieve the explanatory aims of evolutionary biology we need to know of any given biological trait or configuration of the Tree of Life if it is necessary or merely contingent. If necessary, we need to know if it is logically, metaphysically, physically, or biologically necessary. If contingent, we need to know if it is logically, metaphysically, physically or biologically contingent. And similarly for all nonactualised alternatives. Are these nonactualised alternatives nonactual because they are impossible? If so, is that impossibility of a logical, metaphysical, physical or biological nature? If the alternative is nonactual but possible, is that possibility of a logical, metaphysical, physical or biological nature?

I am assuming that these modal categories are related but distinct, and, barring impressive evidence to the contrary, that all categories are occupied. The challenge is to sort actual and alternative effects into these categories. With this task in mind it is easy to see that most contemporary approaches to modality are of little use.

Take the ontological question first: what grounds the modal facts of interest to us? Various answers are on offer. Modal eliminativism, for example, claims the only necessities, possibilities and impossibilities that exist are logical necessities, possibilities and impossibilities, thus doing away with most of our categories. There simply are no facts that allow for the fine-grained distinctions the biologist needs to draw. Obviously such a view offers no assistance because it fails to recognise the full range of modalities the biologist takes seriously.[16]

Modal primitivism, on the other hand, acknowledges non-logical modality, but insists that all non-logical modalities are primitive, non-reducible, non-analysable features of reality, thus debarring herself from offering principled suggestions as to how to sort alternative effects according to their modal character.

Modal conventionalism, i.e., the view that all modality is grounded in our linguistic conventions, is another nonstarter. Biologists will contend, quite rightly, that the extent and configuration of morpho-space have nothing whatsoever to do with our linguistic conventions if only because these are very late arrivals on the biological stage, and it is simply not credible to maintain that the facts which fix the biological modalities had to await the arrival of human language before they could be born.

This leaves the various possible worlds approaches to modality so loved by contemporary metaphysicians. In these approaches what grounds the modal facts are possible worlds construed as (a) distinct, concrete, causally isolated worlds,[17] or (b) abstract objects, usually maximally consistent sets of sentences or propositions[18] or (c) fictional entities.[19] None of these models is attractive from a biological point of view. It is not clear, for one, how states of affairs in causally isolated worlds could be the ontological ground of biological states of affairs in this world. Second, *n'en déplaise à* Plato, it is far from clear how abstract objects of any kind could be the ontological ground of features of concrete entities such as biological forms. And, of course, fictional entities cannot be the ground of any real thing, so *a fortiori* cannot be the ground of biological forms.

Popular contemporary approaches to the epistemology of modality do not inspire much confidence either when considered from a biological point of view. The most widely discussed approaches have sought to develop a connection between what we can *imagine* and our beliefs about modality. For example, conceivability based accounts of the epistemology of modality assert that the conceivability of a state of affairs entails the metaphysical possibility of that state of affairs, or at least counts as evidence for the metaphysical possibility of that state of affairs. An analogous relationship between inconceivability and metaphysical impossibility can also be entertained. When fleshing out just what is involved in conceiving one usually finds the role of the imagination coming to the fore. For example, Yablo argues that p is conceivable for some particular agent if that agent can *imagine* a possible world that would verify p.[20] P is inconceivable if the agent cannot *imagine* a possible world which verifies p. Imagination plays the same role in counterfactual based accounts of the epistemology of non-logical modality. Williamson, for example, maintains that metaphysical modality reduces to logically equivalent subjunctive conditionals, and the epistemology of metaphysical modality is nothing more than the epistemology of counterfactual reasoning. The idea here is that to establish that p is possible one imagines p to obtain and then one runs a mental simulation in which one imagines what else would be the case given p. If nothing contradictory arises in the simulation then p is deemed possible. If the simulation generates a contradiction then p is deemed impossible. As Williamson says, '...our fallible imaginative evaluation of counterfactuals has a conceivability test for possibility and an inconceivability test for impossibility...'.[21]

These accounts should not impress biologists. First, it is not clear that such methods could ever generate the fine-grained distinctions we need to meet our challenge. But the main concern from a biological point of view is the undue confidence philosophers appear to have in the reliability of our powers of imagination to delimit and classify the field of nonactual but possible states of affairs. Most biologists are likely to agree that all biological organs, including the brain, have evolved under selective pressures (although, as we have seen, they need to be careful about this). Adaptive value in the ancestral environment is the primary explanation for a trait spreading in a population, and this applies to belief formation processes as much as any other. But there were no ecological or social factors in play in the ancestral environment that generated selective pressures favouring the emergence of a cognitive capacity that produces true beliefs about *all* possible alternative configurations of the world, particularly those possibilities that have never been actualised, for there are few biological costs involved in having false beliefs about such matters.

The qualification regarding possibilities *that have never been actualised* is particularly important in the present context. It is not unreasonable to assume that the forces of natural selection have so shaped the human mind as to forge *some* positive relation between our powers of imagination and the natural order. Ernst Mach, for instance, maintained that evolutionary forces ensured that humans have 'instinctive knowledge' of the workings of the natural order: 'iron-filings dart towards the magnet in imagination as well as in fact, and, when thrown on a fire, they grow hot in conception as well' (in Sorensen, 1992, 51). And Sorensen, developing Mach, argues that 'events in the world of thought indicate events in the physical world'. He writes:

> ...there are pressures selecting minds that mimic patterns of nature. Hence, the connection between the worlds is biological causation, not an occult bond. Biologists could detect physical patterns of an unexplored planet by studying thought patterns of an alien from that planet. Our prowess for imitation is due to a combination of opposite strategies. The first is impressionability: design a cognizer so it resembles putty. In this way, a large variety of possible regularities can be quickly and precisely impressed upon the mind. The second strategy is hard-wiring. Sometimes we cannot wait for Mother Nature to teach us her lessons. The solution here is to design a creature with innate beliefs. Since the mechanism by which these beliefs are formed is the reliable generate-and-eliminate process detailed by evolutionary biologists, one's innate endowment of truths constitutes a poor man's synthetic a priori. (Sorensen, 1992, pp. 63–64)

There is something to this line of thought, but one should recognise its limitations. The crucial point to bear in mind here is that this account of our reliable knowledge of nonactual possibilities depends on our having had (at least as a species if not personally) experience of the *actualisation* of relevantly similar possibilities. For example, I believe that my house could burn to the ground. I take my belief in this nonactualised possibility to be well founded because I have had experience of relevantly similar houses burning to the ground. This kind of possibility has been actualised many times in the past, and so we are all too aware of this potential hazard. Now the evolutionary argument *does* give reason to think that we can have reliable knowledge of this sort of nonactualised possibility. What it does not do is give particularly good grounds for thinking that our powers of imagination will be as reliable at tracking possibilities that have never been actualised. Many things have proved possible (precisely because they have been actualised) that our ancestors would have deemed inconceivable. But many of the possibilities the biologist is interested in are precisely those that have never been actualised. So while it is undoubtedly true that the ability to imagine *some* kinds of nonactual situations was a tremendous cognitive advantage in the ancestral environment, there is little reason to assume that these advantages presuppose the ability to accurately imagine and thereby identify *all* such possibilities, let alone sort them into the various categories. It is a rationalist conceit to maintain otherwise.[22]

It would appear then that there is little to be had from current approaches to the ontology and epistemology of modality in the way of suggestions as to how one is to determine the modal character of the actual and alternative effects featuring in certain kinds of biological explanations. So let us now consider a radically different approach to these issues stemming from a distinct philosophical tradition.

Scholasticism and modality

To have any hope of reaching a consensus on the sorting of alternative effects in biological explanations according to modal character I suggest we need an ontology and epistemology of modality based on two key assumptions: First, that the modal facts of a given domain are ontologically grounded in the actual features of the entities that make up that domain in this, the actual, world. Second, our epistemology of modality should *not* place emphasis on our powers of imagination. When applied to the biological context a natural first suggestion is that one ought to assume that morphospace is fixed first and foremost by features of *actual* biological forms, and that one gains cognitive access to this morphospace by investigating the empirically available features of these actual forms.

Now the interest of Aristotle and the Scholastics in this context is that they develop an approach to the ontology and epistemology of modality that meets our requirements very nicely. Thus there is a chance, or so I contend, of rectifying the situation outlined in (4) if we avail ourselves imaginatively of the metaphysics of the Scholastics.[23]

Three key Aristotelian theses regarding modality are crucial to what follows. The first is that we have to recognise two distinct kinds of possibility. The first is absolute possibility. In Aristotelian terminology, a *proposition* is absolutely possible if the predicate term is not incompatible with the subject term. This kind of possibility has everything to do with the coherence of our thoughts. But there is a second kind of possibility, viz., possibility relative to some *power*. In this sense something is possible if some power can bring it about. It is this kind of possibility that falls within the purview of the natural sciences, for powers are properties of objects in the natural order. It is relative possibility that is of interest to us in the present context. The second key thesis is that the possibilities that exist in relation to some power are ontologically grounded in essences of the actually existing objects which have the relevant powers.[24] The leading idea here is that it is an entity's nature which sets the boundaries of possibility for it because a nature is ultimately a set of powers and liabilities. A corollary of this is that there is no relative possibility in a domain that is not ontologically grounded in some actuality of that domain. In the biological case, all real biological possibilities are ontologically grounded in the essences of actual forms. The third key claim is that the epistemology of relative possibility reduces to the epistemology of essences. This in turn is largely a matter of determining the causal powers and liabilities of a given object, an enterprise well within the bounds of ordinary scientific investigation.[25]

This approach to modality is attractive as it makes no mention of possible worlds, and places no undue stress on our powers of imagination. It also has the appeal of being fashionable now that the notion of powers has been rediscovered in some contemporary metaphysical circles.[26] But the obvious objection to this line of thought is the well-worn but false claim that evolutionary biology is incompatible with essentialism. But as we saw in the last chapter, this view is based on a misapprehension of the core claims of Aristotle's essentialism and the metaphysical alternatives. Indeed, as we saw, there is good reason to believe that evolutionary biology actually presupposes Aristotelian essentialism.[27] But perhaps the main challenge facing this Aristotelian proposal is the difficulty of identifying biological essences. While many are now happy to recognise the periodic table of elements, say, as specifying the Aristotelian essences of the chemical elements, many are unsure that anything analogous is to be found in biology. But again, matters are not nearly as challenging on this score as they used to be now that geneticists have become comfortable with the notion of developmental programmes.

The development in view here is the process by which complex multicellular organisms are built from single cells. A developmental programme is the set of development control genes and their pattern of expression. Developmental control genes are genes that control the expression of other genes, 'switching' them on and off in a set sequence that terminates in the construction of the mature multicellular organism. It is the developmental programme, rather than the entire genome itself, which determines what sort of organism will ultimately be built from the original single cell.

Now my suggestion in the last chapter was to identify biological essences with species specific developmental programmes that maps genotypes onto phenotypes. This provided a way of accommodating metaphysically evolutionary biology's commitment to the reality of change. Now the same suggestion can be turned to a different use. When this approach to biological essence is adopted the proposal with respect to modality takes the following form:

(1) Potentiality is grounded in actuality, so all strictly biological possibilities must be grounded in some actual feature of this world.
(2) What makes a biological feature biologically necessary, possible or impossible are the essences of currently existing biological forms.
(3) The essence of a biological form is its species specific developmental programme.

(4) X is *biologically* possible if and only if x lies within the scope of the developmental programme of an actual organism. X is *biologically* impossible if it does not fall within the scope of the developmental programme of an actual organism. X is *biologically* necessary if the developmental programme of an actual organism cannot be completed without the occurrence of x.[28]

(5) The other types of modality can then be characterised as follows: X is *physically* possible if it is not ruled out by the causal powers and liabilities of the formal objects of physics and chemistry, *physically* impossible if ruled out by these powers and liabilities and *physically* necessary if it is a consequence of these powers and liabilities. X is *absolutely* or *metaphysically* possible if it is not repugnant to being qua being, *absolutely* or *metaphysically* impossible if it is repugnant to being qua being, *absolutely* or *metaphysically* necessary if it is a consequence of being qua being. X is *logically* possible if it is consistent with the axioms of a given logical system, *logically* impossible if inconsistent with the axioms of that system, *logically* necessary if it is an axiom of that system or follows from the axioms of that system.

(6) The specifically biological modal facts are to be discovered by the empirical investigation of developmental programmes, in particular by intervening in developmental processes.

Some implications of this approach are immediately obvious. First, the sorting of effects into the various biological modal categories will be done primarily by biologists, not metaphysicians. Second, morphospace is not stable, but changes as developmental programmes change. What was formerly biologically possible may become biologically impossible because the relevant developmental programmes are now lost due to extinction. Trilobites are a case in point. Conversely, a current biological impossibility may become possible if the requisite developmental programme should arise. Most importantly, however, what is conceivable or logically possible is not to be confused with biological possibility. And, of course, the physically possible may be biologically impossible. Trilobites, for example, remain physically possible (as well as logically and metaphysically possible) although they are now a biological impossibility. Finally, morphospace is quite constrained – probably larger than some structuralists allow, but much smaller than the neo-Darwinians and Gould have suggested.

However, the crucial point about this approach to modality for present purposes is that it provides a principled method of sorting alternative effects according to their modal character. This in turn provides a way of adjudicating between the various explanations on offer for biological phenomena. If two biologists differ with respect to the explanatory cause they wish to appeal to, in order to explain a given effect, one appealing to natural selection, say, another to physical constraints, we can make a start by sorting the actual and alternative effects according to their modal character to then determine which if any of the favoured causes is a suitable candidate. It is very likely that in the current state of information we will not be in a position to carry out the required sorting. But if we follow the Scholastic's lead, at least we know what sorts of investigations will turn up the needed information. And that is progress.

7
Evolutionary Biology and Ethics

Introduction

In the last chapter an attempt was made to show that Scholastic principles regarding the metaphysics of modality in general and biological possibility in particular can be deployed to help evolutionary biology make good its claim to provide explanations for biological phenomena. In this final chapter I want to consider another aspect of the intelligibility of organisms. The question I have in mind is this: If one knows what possibilities are genuinely open to an organism, can truth-apt determinations be made as to which of these trajectories it would be in the best interests of the organism to realise? Another way of formulating what I take to be essentially the same question is to ask whether biology has anything to contribute to ethics. Can biology be the basis of ethics, as some have supposed? Or is it rather the case that ethics, and moral behaviour generally, are super-biological phenomenon which arrive on the scene when humans somehow rise above their strictly biological nature? These and related questions regarding the relationships that may or may not obtain between biology and ethics have been on the table ever since Darwin's discussion of the origins of morality in chapter 3 of his *Descent of Man*. E. O. Wilson put them front and centre again in 1975 in his (in)famous final chapter of *Sociobiology: The New Synthesis*, and these questions continue to be a source of debate to this day. In this final chapter I want to consider some metaphysical questions arising out of the often tangled relationship of biology and ethics.

An initial puzzle is as follows: It is part of the very self-image of science to be 'value free'. Furthermore, many argue that scientific facts can have no bearing whatsoever on the moral facts (if such there be) because to assume otherwise is to be guilty of the naturalistic fallacy. What is more, if one grants house room to the suggestion that biology *does* speak to ethics, the message seems to be entirely unwelcome, leading as it does to cynicism about the very possibility of ethical behaviour in organisms produced by the processes of natural selection. However, an ancient religious and philosophical assumption is that ethical values are grounded in human nature. This view is gaining currency again in certain circles, and some have sought to employ evolutionary biology to give this approach scientific respectability. Indeed it has been argued in these very pages that evolutionary biology has much to say about human nature inasmuch as the suggestion has been floated that biological essences are species specific developmental programmes. Moreover, biologists themselves, or at the very least those who take themselves to be informed by biology, are often found making statements which appear to go beyond merely reporting on or explaining the facts of biology. Valuing biological diversity per se, and the injunction to preserve habitats and ecosystems, commonplace commitments of conservation biology certainly appear to be moral in character. Indeed there is even an environmental and conservation 'ethics' which insists on the *intrinsic* value of species, habitats and ecosystems.[1] How can this be if science is 'value-free?'

It is necessary to be clear from the outset exactly which questions I am asking regarding the relationship of biology to ethics, for not all such questions are strictly metaphysical in nature. One of the most frequently discussed questions in this area is how evolutionary processes could give rise to organisms capable of moral judgments. Some, Ruse and Wilson (1986), and Ruse (2010) for example, maintain that evolutionary processes associated with kin selection and reciprocal altruism are responsible for the emergence of both our capacity to form moral judgments and for the content of at least some of these moral judgments – both capacity and content are considered adaptations. Others maintain that the capacity for moral judgment is at best an exaptation, and the content of moral judgments has more to do with culture than biology per se (Ayala, 2010). Another related debate concerns the psychological processes involved in arriving at a moral judgment, some suggesting that humans are equipped with a 'moral module', produced by natural selection, that generates immediate responses to circumstances unmediated by rational reflection (Hauser, 2006a, 2006b), while others maintain that rational reflection is a *sine qua non* of a genuine moral judgment (Kant, 1953).

Much is thought to hang on the outcome of these debates. Although sociobiologists have not always been entirely clear on the matter, the suggestion has been floated that the outcomes of these debates will have an impact on our current moral codes, either adding considerations that have previously been lacking, or perhaps challenging current attitudes.[2] And at times it has been suggested that these debates will have an impact on meta-ethical issues. Ruse and Wilson, for example, appear to think that the 'biologicizing' of ethics leads to the conclusion that there is no objective morality.[3] But it is far from clear that anything of metaphysical significance is at play.

The strictly metaphysical questions concerning the relationships between biology and ethics appear to be twofold: First, there is the general location question that can be asked vis-à-vis biological and ethical facts. In the first instance one asks whether biological facts supervene on, or are reducible to, the facts of physics and chemistry; secondly, one asks if the higher-order facts of psychology, say, and possibly ethics, supervene on, or are reducible to, the facts of biology. When asking these location questions we are really asking for an account of how (putative) states of affairs in the natural order hang together ontologically. The second, strictly metaphysical question, is a version of the central question of meta-ethics. Here one asks if biological facts or states of affairs can be the truth makers of moral judgments. The two metaphysical questions are, of course, related; for if one thinks moral or ethical facts are reducible to biological facts, then these biological facts act as truth makers of judgments regarding the reduced ethical or moral facts.

Now the first point to be made here is that the standard debates regarding the evolution of ethics have more to do with the origins of the capacity for making moral judgments, and how moral judgments are arrived at as a matter of psychological fact. But answers to these questions do not in themselves tell for or against the strictly metaphysical questions as I understand them.[4] So, I will ignore considerations related to the origins of our capacity to form moral judgments, as well as those related to the psychological processes involved in the generation of particular moral judgments – unless they are taken to have a direct bearing on the metaphysics of value – and I will focus primarily on the second of the metaphysical questions, namely, whether biological facts can be truth makers of moral judgments (whatever the origin of these judgments happens to be).

Can biological facts be truth makers of moral judgments?

There are many reasons for thinking that biological facts cannot be the truth makers of moral judgments. Some of these have already been touched upon, but it is worth setting them out formally. The following is a sample of what have been taken to be the most pressing considerations:

1. Any attempt to suggest that facts of any description are the truth makers of moral judgments is guilty of the naturalistic fallacy. *A fortiori*, the facts of biology cannot be the truth makers of moral judgements. This argument is deployed by moral anti-realists, who deny the existence of moral facts of any description, and by non-naturalist realists, who deny that moral facts are natural facts.[5]
2. A second line of argument holds that there are no moral facts precisely because the deterministic processes of biology preclude a precondition of morally evaluative behaviour, namely free agency.[6]
3. Even if one were to grant that free agency is not precluded by biology, and countenance the possibility that moral facts are entirely natural, one could still ask whether the facts of biology are compatible with morally praiseworthy behaviour. For, it is argued that morally praiseworthy behaviour is altruistic, and genuine altruism is incompatible with the processes of natural selection. Thus the facts of biology do away with the distinction between good and bad behaviour as all behaviour is morally deficient because it is a product of selfish drives.[7] So if one wishes to continue to draw a distinction between good, bad and better behaviour (a natural assumption of most ethical debates) one will have to assume that the facts in virtue of which these discriminations are made are not biological facts.
4. If the facts of biology do not consign ethics to the dustbin of history as a subject without an object, and if they do not force

the judgment that all morally evaluable behaviour is ethically deficient, there is at least good reason to think that the facts of biology are at best morally neutral. The course of biological evolution shows no sign of moving in the direction of moral progress. There might be a case for saying that evolutionary processes move in the direction of increased diversity and complexity (although even this is not uncontested); but it is far from clear that increased diversity and complexity should be considered moral progress.[8]

5. Even if one grants that there are natural moral facts, it is far from obvious that the facts of biology fix these facts, or could help one identify these facts. One argument for the contrary view is that our moral sensibilities and our capacity to make moral judgments are themselves adaptions. And as adaptations are features of organisms that allow them to cope with their environment more effectively than they would otherwise do, some have thought, arguing from analogy with the sensory modalities, that the value of these adaptations is that they allow us to track moral facts. Now not everyone agrees that our moral capacities and sensibilities are adaptions. Some insist that they are merely exaptations. But even if one were to waive this concern, it is argued that the inference is fallacious. An adaptive belief or belief formation process need not be truth tracking in order to be adaptive. So even if one were to allow that our moral capacities are adaptations, this does not entail that our moral judgments track moral facts.[9]

There is one principal argument to be considered in favour of the positive response to our question.

1. Political theories generally contain a prescriptive component, recommendations regarding our collective actions, institutions and policies. These (differing) prescriptions are based on (differing) conceptions of human nature. As Plamenatz points out in the introduction to his three-volume study of the history of political theory, 'There is always a close connection between a philosopher's conception of what man is – what is peculiar to him, how he is placed in the world – and his doctrines about how man should behave – what he should strive for, and how society should be constructed' (1992, p. xxiv). He insists that this is so '…whether the philosopher is a Rousseau or a Hegel, who does

not agree with Hume that there is no deriving an *ought* from an *is*, or whether he is a Hume himself. For Hume...offers to show how man, being the sort of creature he is, comes to accept certain rules....This way of thinking is not confined to the natural law philosophers and Idealists; it is common to them and to the Utilitarians, and...there is a large dose of it even in Marxism' (1992, p. xxiv). What Plamenatz's historical study shows with respect to political theorists Stevenson illustrates with respect to religious traditions in his *Seven Theories of Human Nature*: '...so much...depends on our view of human nature. The meaning and purpose of human life, what we ought to do, and what we can hope to achieve – all these are fundamentally affected by whatever we think is the "true" or "real" nature of man....Rival beliefs about human nature are typically embodied in various individual ways of life, in different political and ethical systems' (1987, pp. 3–4). The core idea of the tradition which takes human nature to be the touchstone of ethical and political thinking is that the measure of morally evaluable behaviour is human well-being, and that what is good and bad for human beings is determined ultimately by what kind of creatures we are. No one cares to deny that human beings are biological entities. And while it is true that *Homo sapiens* is a cultural animal, our capacity for culture is itself a product of our distinctive biological endowment, in particular our distinctive cognitive evolutionary history. Thus no account of human well-being can be adequate which ignores the facts of human biology. Therefore, the facts of human biology must figure in our moral evaluations of behaviour.

Unless we are simply to ignore the fact that conceptions of human nature figure ineliminably in the most sophisticated political theories developed to date, we must accept that we are faced with a classic aporia, for there are respectable lines of thought on both sides of our question.

The moral facts of biology

Our debate should be considered alongside an additional psychosocial consideration. The sociologist Max Weber famously described modern Western society as 'disenchanted', and he saw this disenchantment as the root cause of the discontents our society is prey to.[10] As disenchantment, and the consequent 'retreat of values from public life', are closely connected to themes under discussion here, it is worth rehearsing his diagnosis briefly in order to place our discussion in the widest possible, and so most revealing, perspective.

Weber's account of our disenchanted society begins with his observation that properly human behaviour is behaviour to which an agent attaches 'meaning'.[11] An action can be meaningful, and also rational, in two distinct ways. First, an action can be meaningful and rational if it serves a purpose. Second, an action can be meaningful if it embodies a value. Weber also argues that human beings need to see their actions and their lives in general as part of a meaningful whole, where meaning is taken in both the instrumental and value-oriented senses. It is this need to see oneself and one's society as part of a meaningful totality that Weber identifies as the primary motivation behind religious and philosophical thought.

Now the distinguishing feature of modern Western society is that the natural sciences have taken over the role previously played by religion and philosophy. This has been a mixed blessing. On the positive side the sciences have been supremely successful in terms of increasing our technological competence and our instrumental rationality. But the downside is that the sciences are taken to militate very strongly against the idea that the universe as a whole, our society, or our individual actions, have any ultimate purpose or value. We are thus left in a disenchanted world, a world in which we know very well how to achieve ultimately pointless ends. The result is the increasing implementation of instrumental rationality in the social, political and economic spheres (the so-called iron cage of rational administration and ever expanding bureaucracy) while we are forced to recognise that the values which fix the ends towards which instrumental rationality is directed are nothing more than subjective attitudes for which no rational justification is possible.

Weber recognises that it is psychologically difficult to live with a worldview that insists on a sharp cleavage between objective facts (as determined by the sciences) and subjective values (deemed to be a product of one's socialisation). His solution – shared with

Nietzsche – that *we* must *give* meaning to our lives, and strive to live by the highest values we find compelling, is of dubious coherence, and is unlikely to get one through the dark nights of the soul. Nor will it do simply to supress the problem. In good times we can throw ourselves into the hustle and bustle of life, and divert ourselves from questions of meaning and value. But in times of personal and social crisis the head-in-the-beta-blockers approach to life proves hollow. And at these moments of stress we are likely to turn on the basic commitments of our worldview, and be tempted to reject precisely those elements of our social order that have played a vital role in our achieving the degree of technological competence we now take for granted.[12] Like Tolstoy, in such moments of discontent many will be tempted to say: 'Science is meaningless because it gives no answer to our question, the only question important for us: What shall we do and how shall we live?'

My point is this: It would be all to the good if some way of bridging the gap between facts and values were to be discovered. If we were able to integrate the sciences into a coherent worldview that would allow one to see one's values as more than the expression of a subjective attitude, this would be a result worth celebrating for the psychological and social benefits it would afford, for one could then see one's actions as not merely instrumentally rational, but rational in virtue of embodying values genuinely worthy of respect. The attempts to ground ethics in theories of human nature can be seen as attempts to provide precisely such a coherent worldview.

In Chapter 3 I mentioned an important feature of the Scholastic commitment to the view that material objects have two metaphysical parts, matter and form. The point was made that by placing the notion of substantial form at the centre of their metaphysical framework the Scholastics are afforded a means of tracing systematic, nonarbitrary and objective relations between metaphysics, epistemology and ethics. The significance of this point here is that the cleavage which renders the fabric of modern Western society is simply not to be found amongst the Scholastics. And this is *not* primarily because the Scholastics were theologians, but because they were Aristotelians. For substantial forms are at one and the same time central to the account of ontologically basic entities (substances such as organisms), central to their account of the intelligibility of substances (forms being one of the four main causes) and, crucial for present purposes, the truth makers of value judgments. The idea is that the form of an organism fixes its essence, thereby shaping what properties the organism has, how it behaves and how it can develop, as well as allowing one to judge whether it is a good, i.e., successful, thriving and healthy, instance of its kind.

Of course one challenge to such a framework is providing an account of the essential forms of things. In the present instance the challenge is identifying the essential form of human beings. If there were such a form, and if this form were to be closely connected to our biology, then biological facts could very well prove to be truth makers of moral judgments. Now I have already suggested in Chapter 5 that there are biological essences, and that biological essences can be regarded as species specific developmental programmes. And in Chapter 6 I suggested that it is precisely such biological essences that delimit the possibilities open to organisms. At issue now is whether these facts of biology can be put to use in the context of value judgments. I suggest they can. What is required is a plausible account of human flourishing or well-being grounded in human biology that can then be used as a skeletal account of the truth makers of value judgments.

Such an account is not difficult to provide if one adopts some assumptions shared by the Ancient Greek ethicists from Plato and Aristotle through to the Hellenistic tradition of the Sceptics, Epicureans and Stoics. The first common assumption was that the study of ethics has as its primary subject matter *not* the rights and wrongs of particular acts, or even the identification of rules according to which one might judge the rights and wrongs of particular acts. At issue first and foremost was to develop an account of the good life for human beings. It is only against the background of such an account that one can begin to consider the merits and demerits of particular actions.[13] Secondly, human well-being, flourishing or thriving in all its forms, was understood to be rooted first and foremost in the notion of *health*. The primary binary opposition in Greek ethical thinking is thus not between Good and Evil, or even between Good and Bad, but between Good and Ill. It is for this reason that much of Greek ethical thinking is cast in the language of health and disease, with ethicists seeking to provide therapy for the ills that plague the human condition.[14] The third point, stressed more by Plato and Aristotle than the Hellenistic schools that followed, is the connection between ethics and politics. As he makes clear in the final chapter of the *Nicomachean Ethics*, Aristotle thought the ills of the human condition susceptible to philosophical therapy could only be treated successfully by the judicious shaping of the laws and political institutions of the society into which we are born. A corollary of this is that ethicists must not confine their attention to the private acts of individuals, but must consider the entire social sphere, including political and economic policy. It is for precisely this reason that the *Nicomachean Ethics* was directed to those who would eventually take on leadership roles within the Athenian state.

Three assumptions thus characterise what I will call the medical conception of ethics: First, that ethics is first and foremost the study of the constitution of eudaimonia for human beings, and how to secure it; second; eudaimonia is rooted in the objective notion of health, somatic and psychological; third, that health in all its forms is determined at least as much by one's political and social context as it is by the private actions of individuals. The suggestion here is that the adoption of this medical conception of ethics proves useful if one's concern is to bridge the gap between fact and value.

A further challenge facing the medical conception of ethics is the provision of an account of human eudaimonia, health or well-being. This is notoriously a topic of endless contention, as it certainly was amongst the ancient Greeks themselves. But the contentious nature of the topic can be overplayed, for while there are no doubt many points on which eudaimonian ethicists disagree, these disputes take place within a context of extensive background agreement. So while it is true that we lack a fully fleshed out view of eudaimonia that commands anything like universal acceptance, there is considerable agreement on some of the necessary features of any human life that could be considered eudaimon. And these uncontentious features of a eudaimon life serve as the brute data against which we measure our ethical judgments, allowing the ethicist to engage in discussions that do not float free of the objective facts of the natural order. These uncontentious features are those adopted by the United Nations' Human Development Index (HDI) as the criteria for measuring social development. And it is no accident that these necessary features of any eudaimon life are closely connected to biological needs. These include:

- Access to adequate nutrition
- Access to potable water
- Access to serviceable clothing
- Access to shelter
- Access to education
- Access to health care
- Security.

Measurable criteria of success include:

- Increase in life expectancy at birth
- Decrease in infant/mother mortality rates
- Decrease in morbidity rates
- Increase in literacy and numeracy rates
- Increase in per capita income (GDP)
- Increase in gender empowerment.

The leading idea here, which can only be waved at in a work devoted primarily to metaphysics, is that the social, political and economic policies of governments or regimes can be graded *objectively* by their record on securing these basic goods for their citizens. If one adopts the medical conception of ethics, the most significant ethical questions concern the *empirical* matter as to which policies are correlated with the highest levels of human development. Once correlations have been established one can then begin the task of tracing cause and effect relationships with the intention of encouraging the adoption of those policies and institutions that have been shown to be most effective in securing the basic requirements of any eudaimon life.[15]

Unless one wishes to contend that health in all its forms is ethically and morally irrelevant, or only becomes so in virtue of our choosing to see them as morally relevant,[16] it should be clear that there is a relationship between biology and ethics, and that it is substantial enough to allow one to employ the facts of human biology to justify morally significant evaluations of social, political and economic policy. For example, if one wants to know whether governments, regimes and parents ought to provide universal education to both girls and boys, the primary consideration is the health, well-being and subsequent development of all concerned, but particularly those who might be denied access to education. The unequivocal answer in the affirmative is supported by the HDI table which shows that all the very highly developed countries have such educational policies, while the less developed countries lag behind in this regard. There is thus very strong *prima facie* evidence that universal education for both sexes positively promotes human well-being, and so there is at least a *prima facie* obligation on governments and regimes, as well as families, to make education a priority. Failing to do so when circumstances permit the implementation of such educational policies is an objective moral failure. Analogous stories can be told for countless other policies. Sen's point that no democracy has ever known a famine is clearly the basis of a moral case for this form of government on the medical conception of ethics. The condemnation of the utterly indefensible practices of female circumcision and the refusal to promote the use of condoms in the prevention of HIV are other immediate consequences of this sort of approach.

Clearly a knowledge of biology alone will not be sufficient to determine on any particular occasion whether a particular policy suggestion is appropriate. Such judgments will have to be reached on the basis of (i) the biologically based features of any eudaimon human life identified earlier; (ii) an appreciation of the kinds of policies and institutions that have shown themselves to be conducive to human development in other contexts;[17] and (iii) an appreciation of the particular circumstances of the society entertaining the policy suggestion. While the goal of eudaimonia remains constant, policies that have worked in one context may not prove successful in others, as the history of developmental economics has shown.[18] These are the pressing empirical questions of ethics that come into view if one adopts the medical conception of ethics. Indeed an implication of this approach is that substantial work in ethics is being carried out in departments of developmental economics and international political economy, areas not standardly covered in courses on ethics offered in philosophy departments.[19] But for present purposes the metaphysical point is that facts of human biology, grounded ultimately in our species specific developmental programme, are the facts against which the merits of policies, practices and institutions are measured.

Replies to the original objections

Let us conclude this chapter with a review of the original objections to the claim that biology might have some light to cast on matters moral. The first concern was that anyone wishing to use biology to justify moral claims is guilty of the naturalistic fallacy. But the naturalistic fallacy makes the mistake of assuming that morally relevant connections between facts and values must be logical or conceptual in nature. The fact that no 'ought' can be derived from an 'is' via acceptable rules of inference can be granted with equanimity once one accepts that there are non-logical connections between facts and values, particularly between health and moral judgments, which must be investigated empirically. One of the lessons of the opening chapter of this book is that metaphysicians should be more than ready to acknowledge the existence and importance of such non-logical relations.

A second concern was that biology and determinism go hand in hand, and as determinism is incompatible with morally evaluable behaviour, taking biology seriously means either giving up on ethics altogether, or confining ethics to a realm of fact, presumably super-biological, in which freedom from the laws of nature allows for the possibility of moral responsibility. But this line of thought is attackable on a number of fronts. First, Frankfurt has argued that moral responsibility is in fact compatible with determinism, so one might accept the first assumption of the argument but reject the second. But one might also reasonably baulk at the first. It is not merely that one's genes do not determine one's phenotype in anything like as rigid as fashion as previously thought; there is nothing in one's genes to force the metaphysically significant thesis of actualism, i.e., the view that what *actually* happens is all that ever *could have* happened. In fact recent work on the evolution of human cognition suggests that one important selective pressure was precisely the need to widen the behavioural repertoire of human beings to accommodate the fact that our social and ecological environment is significantly more complex than that occupied by other animals. We are free agents precisely because of our biological endowment.[20]

Another concern was that for an action to be deemed 'moral' it has to be altruistic, and it is impossible for genuine altruism to be produced by the forces of natural selection. It is this line of thought that leads to Darwinian cynicism about the very possibility of morally acceptable behaviour. We can set aside the biological facts in this case, for this argument pivots on a dubious idea of what constitutes a moral act. On any respectable account of ethical behaviour it is indeed the case that the interests of those affected by the actions of an agent have to be given due consideration. But this is a far cry from insisting that a moral act must be to the disadvantage of the agent, or at best neutral, and only beneficial to those affected by it. This conception of the moral is not forced, and is certainly not a component of the medical conception of ethics. A minimal requirement of an act counting as moral is that the agent is not the only beneficiary of the act. It is not required that the agent act against her own interests. There is nothing in biology that militates against such a possibility.

The replies to the final two arguments are obvious enough. Nothing in the case offered here turns on the idea that evolution is moving in the direction of progress, moral or otherwise. And no use is made of the idea that our moral capacities and sensibilities might be adaptations. Instead the argument hangs on the idea that thriving, or failing to thrive, while always matters of degree, are ultimately matters of biological fact. It is to the credit of the Scholastics that their metaphysical principles allow this fact to be clearly appreciated. But it is simply another instance of Scholastic principles taking on an unexpected lustre when metaphysics is approached from a biological point of view.

Notes

Introduction

1. In Moore (1963).
2. See Oderberg's (2007) as well as the recent collections of Feser (2013) and Takho (2012).
3. The degree of sophistication reached by scholastic philosophers has been noted (if not always appreciated) even by some social anthropologists. Writing of the division of labour to be found in agrarian societies, Gellner writes: 'Some of the specialisms of a mature agrarian society will be extreme: they will be the fruits of life-long, very prolonged and totally dedicated training, which may have commenced in early youth and required an almost complete renunciation of other concerns. The achievements of craft and art production in these societies ... often reach levels of intricacy and perfection never remotely equalled by anything later attained by industrial societies, whose domestic arts and decorations, gastronomy, tools and adornments are notoriously shoddy. Notwithstanding their aridity and sterility, the scholastic and ritual complexity mastered by the schoolmen ... is often such as to strain the very limits of the human mind (*Nations and Nationalism*, p. 26)'. Comments on aridity and sterility apart, those acquainted with Scholastic texts will wholeheartedly agree with Gellner's observations.

1   A Return to Scholastic Metaphysics

1. See Nietzsche's (1966) *Beyond Good and Evil*, in particular chapter 1 'On the Prejudices of the Philosophers'. He writes: '[M]ost of the conscious thinking of a philosopher is secretly guided and forced into certain channels by his instincts. ... They all pose as if they had discovered and reached their real opinions through the self-development of a cold, pure, divinely unconcerned dialectic ... I do not believe that a "drive to knowledge" is the father of philosophy; but rather that another drive has, here as elsewhere, employed understanding (and misunderstanding) as a mere instrument' (pp. 11–13).
2. The later, therapeutic, Wittgenstein, is perhaps the most prominent exponent of this view.
3. I take Carnap's "Empiricism, Semantics and Ontology" (1950) and Quine's "On What There Is" in his (1980) to be the classic exemplars of this position.

4. In fact Gilson calls this the 'first law to be inferred from philosophical experience' (1937, p. 306).
5. Chalmers, Manley and Wasserman's (2009) collection of essays on meta-metaphysics provides a good indication of the wide range of competing views currently entertained on methodological issues.
6. Hobbes never had a good thing to say about the Scholastics, and caustic remarks are to be found throughout his works. These are just a selection taken from *Leviathan*, chapter 46, 'Of Darkness from Vain Philosophy and Fabulous Traditions'.
7. With a few notable exceptions. See Leibniz' *Discourse on Metaphysics*, section 11, where he writes: '...our modern philosophers do not do enough justice to St. Thomas and to other great men of that era...the views of the Scholastic philosophers and theologians have much more soundness than is imagined' (1983, p. 22).
8. If one is not entirely comfortable with this form of words (perhaps because one is not entirely comfortable with the Quinean criterion of ontological commitment and the use to which Fregian logic is frequently put in metaphysical discussions) one can characterise the domain in question as being what the Scholastics referred to as 'real' being. As Suarez puts it: 'real being' embraces 'everything that is not absolutely nothing [...] that can exist in reality otherwise than as a figment of the intellect' (2007, p. 37). Real being is to be contrasted with 'beings of reason' (*entia rationis*), examples of which include privations, negations and chimera, as well as genus, species, antecedents and consequents. For a good recent study of Suarez on beings of reason see Christopher Shields' 'Shadows of Beings: Francisco Suarez's Entia Rationis' in Hill and Lagerlund's (2012).
9. This aspect of metaphysics focuses on 'how things hang together' as an organised system of entities, with particular emphasis on the question of ontological dependency.
10. Note that this is a far cry from the view, found in Kant (1965, pp. 9–11), that metaphysics is the attempt to arrive at knowledge of reality 'independently of all experience'. Just why Kant makes such claims, while an Aristotle does not, is part of the transition that is the subject of this chapter.
11. I am using the 'formal'/'material' distinction here in the traditional sense. One and the same object can be considered from different points of view. The different points of view generate different *formal* objects despite the fact that one and the same *material* object is the subject of both contemplative acts. The formal object/material object distinction is part of a mindset found in Aristotle according to which what distinguishes metaphysics from physics and mathematics is not their material objects (for these can be shared), but rather their formal objects. Metaphysicians, physicists and mathematicians confront one and the same world, but focus on different aspects of it, thereby producing different formal objects. For an extended discussion of this approach

to the division of the various sciences and their interrelationships see Aquinas, *Commentary on the* De Trinitate *of Boethius*, q. 5, in his (1986).

12. The emphasis on the necessary is already present in Plato's *Timaeus*. In this historically most influential of the dialogues, Plato provides his account of the origins and nature of the universe. And in one passage he speaks as follows: 'All these things [i.e., the elements out of which the universe is made] being so constituted of necessity, were taken over by the demiurge. ... We must accordingly distinguish two kinds of *aitia* (causes), the necessary and the divine. The divine we should search out in all things for the sake of a life of such happiness as our nature admits; the necessity for the sake of the divine, reflecting that apart from the necessary ones the divine ones, which are the only objects of our serious study, cannot be apprehended or grasped nor can we participate in them in any other way (68e1–69a5).

13. *Metaphysics*, Book VI, chapter 2, in (1941). Translation by W.D. Ross.

14. It is worth pointing out here that the tight connection between objective necessity and subjective certainty is found not simply amongst philosophers but includes statisticians as well, as is revealed particularly in the notion of probability. Consider the conflation of necessity and certainty in the following passage taken from an introduction to statistics: 'The probability calculus assigns numbers between 0 and 1 to uncertain events to represent the probability that they will happen. A probability of 1 means that the event is *certain* (e.g., the probability that, if someone looked through my study window while I was writing this book, they would have seen me seated at my desk.) A probability of 0 means that an event is *impossible* (e.g., the probability that someone will run a marathon in ten minutes). For an event that *can* happen but is neither certain or impossible, a number between 0 and 1 represents its "probability" of happening' (Hand, 2008, p. 57). This passage reveals how natural it is to slide between talk of the necessary, possible/contingent and impossible on the one hand (objective modalities), and talk of epistemic certainty and uncertainty on the other. Because it is assumed that there can be certainty about the necessary and the impossible, with certainty following on from an appreciation of the modalities of a situation, but only uncertainty regarding the possible/contingent, one can often use the term 'certain' when one ought in all strictness to say 'necessary'. With this in mind consider the following passage: '...a throw of the die might produce a 1, 2, 3, 4, 5, or 6 and the symmetry of the die suggests that the six outcomes are equally likely, so each has a probability of 1/6 (they *must* sum to 1, since it is *certain* that one of 1, 2, 3, 4, 5, or 6 will come up)' (Hand, p. 59). It is certain because no other result is possible.

15. *Commentary on the De Trinitate of Boethius*, q. 5, a.1, p. 13, in (1986).

16. In Kirk, Raven and Schofield (1991, pp. 193 & 419).

17. Translation by Benjamin Jowett.

18. In Bosley and Tweedale (2006, pp. 37–40).

19. For a good discussion of the nature of this theological task see chapter 1, 'From Story to System', in Rosemann's (2004).
20. These are taken from Klima (2007, pp. 182–184).
21. Aquinas would prefer one to say that God's omnipotence amounts to the ability to do anything that is possible. If God cannot bring about a logical contradiction it is not because of any incapacity in His power but because of the incapacity on the part of the logically impossible to be brought into existence (see *Summa Theologica*, I, q. 25, a. 3).
22. It is highly significant that in this same work he writes: 'I declare that, neither in this treatise nor in others, do I wish to say anything which is against the articles of faith, or against the decision of the Church, or against the articles the opposite of which were condemned at Paris' (ibid., 33).
23. This is a constant refrain in Ockham, and is usually grounded on the claim that 'I believe in God the Father Almighty', the opening line of the Apostle's Creed. See for example the *Fourth Quodlibet, Question 22* in his 1991.
24. See Ockham's 1991 *Fourth Quodlibet, Question 22*. An implication of this claim is that God can cause one to perceive a tree, say, without the perception being caused by a tree. It is striking that Ockham uses the term 'demon' in an argument for scepticism based on the consideration expressed in (2). See his *Ordinatio* I.27.3, in Pasnau (2002, p. 235).
25. See Ockham's 1991, *Fifth Quodlibet, Question 18*. This point was particularly sensitive in the context of the Eucharist. In Ockham's view the bread and wine can become the flesh and blood of Christ, making Christ literally present despite the fact that one continues to see bread and wine. This is metaphysically possible because there is no contradiction in the claim that the underlying substances can change while the accidents do not.
26. See his *Reportatio, II, Question 150*.
27. This summary of Ockham's fundamental assumptions is taken from Ockham (1990).
28. Nicolas does not necessarily endorse them, however. In fact he is explicitly asking for advice on how to avoid these consequences which he thinks inevitably follow from positions sanctioned by the 1277 condemnations. Nicolas might be drawing out the implications of the condemnations of 1277 with a view to exploiting these consequences as a *reductio* of positions popular among his contemporaries in Paris. See Zupka (1993, pp. 191–221).
29. In 1330–1331 Crathorn took a similar line as Ockham and Autrecourt. In his *Questions on the First Book of Lombard's Sentences*, Q.I, he argues that because the accidents of the consecrated host are indistinguishable from those of unconsecrated bread, one cannot tell on the basis of sense experience whether one's visual experience of bread is caused by bread, or the body of Christ, or, indeed, nothing at all. In Pasnau (2002, p. 290).

30. The connection between the condemnations of 1277 and Cartesian scepticism has been noted before but the importance of the point has not been recognised. See Bermudez (2000) and Frede (1988).
31. See the preface to Pinborg's (1976) and M. Markowski (1984) for able discussions of this point.
32. Consider also Spinoza's remark in his *Principles of Cartesian Philosophy* regarding the origins of the hypothesis. He writes: 'And there was a particularly strong reason, *an ancient belief*, fixed in his mind, that there was an all-powerful God who had created him, and so may have caused him to be deceived even regarding those things that seemed very clear to him. This then is the manner in which he called everything into doubt' (1998, p. 8).
33. This is what Descartes says of Francis Suarez in the course of his fourth set of replies (2008, p. 164).
34. There are no such doubts in the case of Leibniz. He claims to have read Suarez in his youth.
35. It is more than possible that the explanation has something to do with the fact that Descartes is even more extreme than I have portrayed him here in that he denies even the existence of logical necessity. However, it is difficult to reconcile Descartes' view that the only necessity is conformity to the will of God with his, or any, philosophical practice.
36. Consider also Leibniz's claim that, ultimately, there is no hard and fast distinction between necessary and contingent truths, for *all* affirmative truths are knowable a priori, at least by God, because they are ultimately *analytic* truths. See his essay 'Necessary and Contingent Truths', the opening lines of which read: 'An affirmative truth is one whose predicate is in the subject: and so in every true affirmative proposition, necessary or contingent, universal or particular, the notion of the predicate is in some way contained in the notion of the subject, in such a way that if anyone were to understand perfectly each of the two notions just as God understands it, he would by that very fact perceive that the predicate is in the subject'(1983, p. 96).
37. This point is recognised by Gendler and Hawthorne (2002).
38. See *The World*, in Descartes' (1990, p. 89).
39. However, in defence of the lingering effects of (1) and (2), many epistemologists are still at work on what has become known as the 'traditional epistemological project' – the project, to use Carnap's terminology, of constructing the world of physics, other minds and cultural objects on the basis of one's 'autopsychological' experiences.
40. A very early version of this argument can be found as far back as Joseph Priestly's *Disquisitions Relating to Matter and Spirit*. See Yoltan (1983, p. 114).
41. (1989) *A Treatise of Human Nature*. Both passages are from Book I, Part III, section XIV.
42. On the first point he says: 'Experience teaches us that a thing is so and so, but not that it cannot be otherwise. First, then, if we have a proposition

which in being thought is thought as *necessary*, it is an *a priori* judgment' (1965, p. 44). On the second we find the following: 'We cannot determine from mere concepts how...because something is, something else must be, and how, therefore, a thing can be a cause ...' (ibid., p. 253).

43. This soon changes after 1277. In 'Medieval Thought Experiments: The Metamethodology of Medieval Science' Peter King shows how thought experiments become important in medieval science in the mid-fourteenth century, i.e., after 1277. In Massey and Horowitz (1991, pp. 43–64).

2 The Aporetic Method and the Defence of Immodest Metaphysics

1. I have in mind here, of course, the specifically Quinean recommendation of determining one's ontological commitments by expressing an accepted scientific theory in the language of first-order predicate logic and then committing oneself to those entities that must be admitted as possible values of bound variables if the theory is deemed true – making due allowances for possible recourse to paraphrases. Here the heavy lifting is done by the scientific theory. Conceptual analysis has a similarly deflationary impact on the role of the metaphysician, and appears to provide only 'trivial' or 'internal' answers to metaphysical questions. Conceptual analysis is used to clarify the application conditions of key concepts, thereby clarifying the truth conditions of existence claims containing those terms and the entailment relations between sentences containing those terms and those which do not. With such analyses in hand it is very often easy to determine whether the application conditions of a relevant concept obtain or not by reference to uncontroversial first-order empirical knowledge or other uncontested claims. Using this approach one can arrive very quickly at apparently metaphysically heavyweight claims. To take a current example, one can claim that abstract objects like numbers exist because an uncontroversial statement like 'The number of fingers on my hand is finite' entails the existence of numbers. But again, this approach does not leave much for the metaphysician *qua* metaphysician to do.

2. I agree with Sider that most contemporary forms of ontological antirealism stem from concerns relating to epistemology. He suggests that current forms of anti-realism are 'based on the desire to make unanswerable questions go away'. These questions are deemed unanswerable because they 'resist direct empirical methods but are nevertheless not answerable by conceptual analysis' (In Chalmers, Manley and Wasserman, 2009, p. 419).

3. By contrast with Hofweber's *modest* metaphysics. See his (2009).

4. Standard variations on this final theme emphasise the possibility that parties to an ontological dispute are talking past each other because they attribute different senses to key terms (the existential quantifier and the

term 'object' being prime candidates). It is also suggested that perhaps ontological claims are not statements all, but merely prescriptions about how one should talk. One might also suspect that a subject or predicate term in an ontological statement is itself problematic in some fashion, rendering it impossible to attribute truth conditions to the statements in which it is embedded.

5. One might also be tempted to reject metaphysics on the grounds that it is an undesirable activity (perhaps because male-biased, exclusionary and oppressive). This postmodernist and feminist critique deserves close consideration; but as it does not address the *viability* of metaphysics as a distinct study, I will set it aside here. But see Charlotte Witt (in Van Invagen and Zimmerman, 1998) for an argument to the effect that even feminist philosophers must engage in metaphysical reflection as traditionally understood in order to forward their agenda of social reform.

6. This paragraph is heavily indebted to Lowe (2001, pp. 3–8).

7. In actual practice many argue as though a hierarchy of scientific authority does exist, and so physics is given precedence over chemistry and biology, and the latter are privileged over, say, linguistics and psychology. But it is not clear that this hierarchy can be justified without appeal to some implicit metaphysical principles.

8. This argument is due to Lowe (2001, p. 5). See also his (2007, p. 5). This is in line with a standard view that the sciences deal with the actual, while metaphysics deals with the necessary and the possible.

9. See in particular Thomas Kuhn's 'Second Thoughts on Paradigms' in Suppe (1974).

10. See Duhem's *The Aim and Structure of Physical Theory* for a perceptive and detailed account of scientific theory shorn of metaphysical baggage. A key point of Duhem's masterpiece is that the price of eschewing metaphysical commitments in physics itself is the adoption of an anti-realist attitude vis-à-vis the theories of physics.

11. This is a point of no small moment at a time of financial austerity within higher education.

12. Not all great philosophers fall into this camp, but it is clear that Plato, Aristotle, the late Hellenistic thinkers, the Scholastics, Descartes, Spinoza, Hume and Kant, not insignificant figures, do. All were concerned not just with the theoretical aspects of the discipline, but with its practical consequences as well.

13. Again, this view is hardly original. The standard view that philosophy is not a first-order discipline is again nicely expressed by Sellars: 'Philosophy in an important sense has no special subject-matter which stands to it as other subject-matters stand to other special disciplines.... What is characteristic of philosophy is not a special subject-matter, but the aim of knowing one's way around with respect to the subject-matters of all the special disciplines'. ('Philosophy and the Scientific Image of Man' in *Frontiers of Science and Philosophy*. Robert Colodny, ed. Pittsburgh: University of Pittsburgh Press, 1962).

14. Aristotle stated that an aporia arises ' ... when we reason on both sides [of a question] and it appears to us that everything can come about either way'. This produces 'a state of *aporia* about which of the two ways to take up' (*Topics*, VI, 145b16–20).

15. This view of the nature of what I am calling coordination problems is found in Ryle's *Dilemmas*. In chapter 1 of this work he writes: 'Certain sorts of theoretical disputes, such as those we are to consider, are to be settled not by any internal corroboration of those positions, but by an arbitration of quite a different kind – not, for example, to put my cards on the table, by additional scientific researches, but by philosophical inquiry' (p. 5).

16. These points draw heavily on the following extended passage from Aristotle's *Metaphysics*: 'We must, with a view to the science [metaphysics] which we are seeking, first recount the subjects that should be first discussed. [The "subjects" being a set of aporia.] ... For those who wish to get clear of difficulties [i.e., aporia] it is advantageous to discuss the difficulties well; for the subsequent free play of thought implies the solution of the previous difficulties, and it is not possible to untie a knot of which one does not know. But the difficulty of our thinking points to a "knot" in the object: for in so far as our thought is in difficulties, it is in like case with those who are bound; for in either case it is impossible to move forward. Hence one should have surveyed all the difficulties beforehand, both for the purposes we have stated and because people who inquire without first stating the difficulties are like those who do not know where they have to go; besides, a man does not otherwise know even whether he has at any given time found what he is looking for or not; for the end is not clear to such a man, while to him who has first discussed the difficulties it is clear' (995a23–995b4).

17. It is a common though unfortunate fact that philosophical discussions often lose sight of the original problem that gave rise to the discussion in the first place, lending a curiously detached air to the proceedings.

18. This approach has much in common with that of the early analytic philosophers who hoped that all philosophical questions arise from conceptual confusions. Once these confusions have been identified 'everything', i.e., the lines of thought leading to the aporia, 'will be left as it was'.

19. For further discussion and examples see Nicolai Harrmann's (1965); my (2007, chapter 1); and Rescher's (2006, 2009).

20. Gary Gutting (2009) has recently argued that the philosopher's contribution to the general pool of knowledge is precisely the knowledge of distinctions. See also chapter 4 of Rescher's (2006).

21. I have in mind such conflicts as have arisen between the competing interpretations of Quantum Mechanics; between Quantum Mechanics and Relativity Theory; between Quantum Chemistry and Standard Chemistry; between Chomsky's linguistics and evolutionary biology; between neuroscience and psychology, to mention only a few. But not all inter-scientific conflicts will require recourse to metaphysics. The

conflict that arose between genetics and systematics prior to the neo-Darwinian synthesis, for example, I would be inclined to discount as an aporia calling for metaphysics treatment. The great neo-Darwinian synthesis was brought about by a close examination of the available data. On the standard interpretation of the historical background to the synthesis, systematists and naturalists believed that genetics forced the adoption of saltationism rather than the gradualism they preferred. This was a misapprehension born of the fact that most naturalists were not aware of the latest discoveries in genetics which had established that mutations were common enough and small enough to allow for gradualism. On the other hand, geneticists knew how to handle adaptations, but could not explain the origin of biological diversity (speciation). But unbeknownst to them this gap had already been filled by the systematists and naturalists who had shown that the key to the problem of speciation was isolation. The brilliance of Dobzhansky (1949) and Mayr (1942) lay in showing that the available data from the two branches of biology were in fact compatible; and this was achieved in large part by bringing the various branches of the discipline 'up to speed' with the latest data from other areas.

22. Space considerations make detailed discussion of alternatives impossible, but some comment, however brief, is appropriate. Two approaches have been mooted in the literature (taking it as read that an ontology cannot be read off in any straightforward fashion from the sciences). The first, and by far the most widely accepted, is to take ontological claims as quasi-scientific hypotheses to be assessed by a battery of criteria such as explanatory power, simplicity, coherence with other theoretical commitments, common sense or 'intuitions', etc. But there are several shortcomings with this tack. First, the criteria rarely all pull in the same direction, and there is no obvious way in which they can be ordered in terms of authority. Moreover, views differ on the second-order criteria used to judge whether a theory is, for example, simpler or more economical than another. The upshot is that the battery of criteria approach, while lending some degree of warrant to an ontological claim, rarely provides anything more than fig-leaf rationality as just about *any* ontological claim can be warranted on some construal of these criteria. This comes dangerously close to conceding to the sceptic that *no* ontological claim is ever more warranted than any other.A second approach to the justification of ontological claims is due to Lowe (2001, 2006). One begins by developing a set of ontological categories knowable a priori and arrived at via rational debate, the contention being that these categories fix the metaphysical possibilities. The mature sciences are then called upon to determine empirically which of these possibilities have been actualised. Finally, these actualised possibilities, discovered thanks to the combined efforts of metaphysician and scientist, are taken to be the fundamental structure of reality. This approach has the attraction of recognising a division of labour between the sciences and metaphysics; but it suffers for want of detail regarding the crucial first step. How, exactly, will

rational debate produce the required set of categories? Lowe leaves this most important of questions without an explicit answer as far as I can see. If, as one suspects, it depends ultimately on the battery of criteria approach, then it inherits all the attendant defects of that approach. If it relies on intuitions, then it falls foul of epistemic naturalism.

23. This paragraph is heavily indebted to John Stachel's work on the origins of the Special Theory of Relativity (2002).

24. See Endnote 23.

25. Irwin (2002, pp. 8–10), *Aristotle's First Principles*. Oxford: Clarendon Press. A closely related complaint is that the method cannot generate a warrant for full-blooded metaphysically realist claims, although those of the anaemic 'internal' realist variety might gain some support.

3   Evolutionary Biology Meets Scholastic Metaphysics

1. Clearly in a work of this length I cannot consider all biological aporia, nor can I consider aporia arising out of other sciences. So no complete vindication of Scholastic metaphysical principles is possible here. However, as biology, and evolutionary biology in particular, have often been thought to be incompatible with Aristotelian principles, it would be significant if these principles should hold up under even this limited scrutiny.

2. For further discussion see chapter 2 of Sterelny and Griffiths, *Sex and Death: An Introduction to Philosophy of Biology* (Chicago: University of Chicago Press, 1999).

3. Mayr writes: 'Philosophers have endlessly speculated about whether there is a real world outside of us... and whether this world is exactly as we are told by our sense organs and by science.... Biologists known to me are commonsense realists' (1998, p. 56). 'Common sense is not a fashionable tool among philosophers, who much prefer to rely on logic.... [But] a commonsense approach is often the most comfortable and productive ...' (ibid., p. 57).

4. It is worth noting that it is not entirely clear just what is meant by the claim that some objects are abstract or universal as there is still no settled view as to the correct characterisation of the distinction between particulars/singulars and abstract objects/universals, despite the importance of the distinction in both metaphysics and epistemology. Perhaps the most common view is to say that abstract objects are non-spatiotemporal, and causally inert, while all singulars exist in space and time and enter into causal relationships with other singulars. Lowe (2002, p. 351) has suggested that singulars can be characterised as entities that are instances of other entities while having no instances themselves, whereas universals can both be instances of other entities while being instantiated themselves. Zalta (1983, p. 60) has also suggested that abstract objects might be viewed as possibilia. But for present purposes it is simply worth putting down a marker that it is not entirely clear what the denial of (5) amounts to.

5. I say 'metaphysical part' advisedly, for this is perhaps the key distinguishing feature of the Scholastic framework. The Aristotelian position on immanent form distinguishes it from the metaphysical systems of Plato and Democritus, as well as those of the early modern period which can be characterised as a group as those which insist that material objects have only integral parts and no metaphysical parts. Aristotle's rejection of the atomism of Democritus, and the Scholastic rejection of corpuscularism in the Early Modern period, is made on precisely the point that one cannot give an adequate account of a material object by adverting solely to its integral parts.

6. For a very clear statement of this position see Suarez, *Metaphysical Disputation VI*, section 5 (1964, pp. 70–73).

7. In virtue of this tendency singulars were also called universals, but only by extensive denomination and emphatically not in the same sense in which concepts are called universals.

8. This process is one scientists are quite familiar with. Duhem's account of the role of physical laws in scientific thought is very close to the Scholastics' parallel account of concepts. Duhem writes: 'The human mind had been facing an enormous number of concrete facts; no man could have embraced and retained a knowledge of all these facts; none could have communicated this knowledge to his fellows. Abstraction entered the scene. It brought about the removal of everything private or individual from these facts, extracting them from their total only what was general in them or common to them, and in place of this cumbersome mass of facts it has substituted a single proposition, occupying little of one's memory and easy to convey through instruction: it had formulated a physical law' (1977, p. 22).

9. Even Scotus accepts this account as far as it goes. However, he notoriously goes further than an Aquinas or Ockham by insisting that there is an extra-mental common nature which is not really but only formally distinct from singulars. But even he does not suggest that extra-mental reality contains universals in the strict sense. This may come as a surprise to many who believe that late Scholastic nominalism and Scholastic realism were fundamentally at odds on ontological matters. However, as Klima has shown in his introduction to Buridan's (2001, liii–lxi) (2001), the dispute between these schools was primarily over questions of semantics and not ontology.

10. I have taken the formulation (if not the numbering) of many of these principles from Wuellner's far more comprehensive list to be found in his (1956).

4 Counting Biological Individuals

1. See Jack Wilson's (1999) for a recent book length treatment of biological individuality that begins by recognising the practical difficulty. For early discussions of the concept of the biological individual see T. H.

Huxley's (1852), J. Huxley's (1912) and E. Haechel's (1979). For recent work suggesting that a grasp of the notion of the biological individual is absolutely essential to the viability of evolutionary theory, and that such a grasp is currently lacking, see Pepper and Heron (2008) and Ruiz-Mirazo et al. (2000).

2. For further discussion see E. F. Keller, 'It Is Possible to Reduce Biological Explanations to Explanations in Chemistry and/or Physics', in *Contemporary Debates in Philosophy of Biology*. Ayala and Arp (eds), (Oxford: Wiley-Blackwell, 2010, pp. 19–31), and John Dupre, 'It Is not Possible to Reduce Biological Explanations to Explanations in Chemistry and/or Physics', in *Contemporary Debates in Philosophy of Biology*. Ayala and Arp (eds), (Oxford: Wiley-Blackwell, 2010, pp. 32–47).

3. As Hull says, 'If selection is a process of differential perpetuation of the units of selection, and if organisms are the primary focus of selection, then we'd better know which entities we are to count' (2001, p. 17). Of course there are those who champion the gene as the basic unit of selection, and others who, while recognising the individual organism as one unit of selection, wish to add others such as the group or species. But both accept that the organism plays a key role in any account of selection. Groups are obviously made up of individual organisms, and genes are subject to selection indirectly through direct pressures brought to bear upon their interactors, i.e., the organism they use to replicate themselves.

4. 'What the ecologist, as population biologist, studies is the density of a population (number of individuals per unit area), the rate of increase (or decrease) of such a population under varying conditions, and, when dealing with the populations of a single species, all those parameters that control the size of the population such as birth rate, life expectancy, mortality, and so on' (Ernst Mayr, 1997, pp. 210–211).

5. We are standardly told that '...the fitness of a trait depends on the proportion of the individuals in a population that have that trait' (Sterelny and Griffiths, 1999, p. 8). Similarly, '...difference in survival is expressed...by a parameter, assigned to each genotype, called *relative fitness*, a number that represents the fraction of the individuals of a given genotype that survives to adulthood and reproduction' (Stearns and Hoekstra, 2005, p. 76).

6. 'In contemporary evolutionary biology an "adaptation" is a characteristic of an organism whose form is a result of selection in a particular functional context' (West-Eberhard, 1998, p. 8).

7. The need to ensure that we are comparing 'apples with apples' when employing the comparative method is part of the case Pepper and Heron (2008) make for the indispensability of an account of biological individuality to evolutionary theory.

8. This paragraph is heavily indebted to Wilson (1999, pp. 6–7).

9. For extended discussion of the attending difficulties of these criteria see Ellen Clarke (2010).

10. The first two chapters of Strawson's *Individuals* is a classic discussion of individuation in this sense.

11. This is how the problem is characterised by Lowe in his (2005, pp. 75, 79).
12. Amongst the Scholastics there was a consensus on the point that everything insofar as it exists is an individual, although there was ongoing debate as to whether the properties of substances were individuals in their own right in addition to substances themselves (see *Metaphysical Disputation* V, section I).
13. Suarez was very much aware of this complication. In response to an objection to his standard answer to the constitutive question he writes: 'But you will say: At least this water will not be singular, because it is divisible into many [entities] in [each of] which the whole nature of water is to be found. The answer [to this] is that what is divided [this water] is not divisible into many [entities] which are also "this water", but [rather] which are [just water]; and hence "this water" is singular, while "water" is common'. (*Metaphysical Disputation* V, section I, p. 32.)
14. A typical example of the principle at work can be found in *Summa Theologiae* (1988) I, q. 75, a 2. 'Nihil autem potest per se operari, nisi quod per se subsistit. Non enim est operari nisi entis in actu: unde eo modo aliquid operator, quo est'. 'Now only that which subsists in itself can have an operation in itself. For nothing can operate but what is actual, and so a thing operates according as it is'. English translation by Pegis (1997). This is reaffirmed in *Quaestio Disputata de Anima*: 'everything that exists in its own right has its own activity' (in Aquinas, 2008, p. 184).
15. The plausibility of linking individuals to actions and operations can be reenforced by other considerations as well. (1) The focus on action and operation is itself sensible because we do not live in a frozen world. Because change is a real feature of the world we inhabit, and change is primarily brought about by the actions of individuals, it is no surprise that our notion of the individual per se is closely connected to that of action. (2) Linking individuals to actions also delivers a common sense response to the discussion of mereological sums. There is good reason to consider an apple, say, to be a real entity above and beyond the atoms that make it up because there are actions associated with the apple that cannot be attributed to the atoms in and of themselves. Similarly, we do not usually consider an apple and a pear to constitute three objects (the apple, the pear and the apple/pear composite) but only two because there are no actions associated with the apple/pear composite. A corollary of *agere sequitur esse* is the principle that where there is no action, there is no being. (3) Tying individuals to action also allows one to sidestep Unger's sorites paradox of composition. Focusing on the fact that a biological individual is ultimately a specific kind of agent rather than a collection of cells allows one to deny the second claim of his aporia. It also allows one to recognise the difference between a living organism and a fresh corps.
16. Mitochondria lost the capacity to photosynthesize, while chloroplasts lost respiration.

17. See Stearns and Hoesktra, chapter 15, for extended discussion.

5 Evolutionary Biology, Change and Essentialism

1. This is what I take Lewis to be doing in his discussion of the problem of temporary intrinsic properties (1987, pp. 202–205). To allow for 'change' in intrinsic properties he has to say that 'different intrinsic proper-ties…belong to different things.…There is no problem at all about how different things can differ in their intrinsic properties' (p. 204). This is undoubtedly true, but the point about change is that *one and the same thing* is to have different intrinsic properties at different times, not two distinct things. Hence the assertion that perdurance theories in effect deny the reality of change.
2. See Lowe (2002, chapter 3) for a critical discussion of perdurance and temporal parts theories.
3. See my (2007) for an extended defence of this Moorean approach to conflicts arising between common sense and philosophy.
4. This aporia exists for anyone who takes evolutionary biology and contem-porary metaphysics seriously. Few scientific theories enjoy the prestige of evolutionary biology. Stearns and Hoekstra rightly insist that 'The ideas of evolution have survived many controversies and tests and are now considered as reliable as any ideas in science'. *Evolution: An Introduction* (Oxford: Oxford University, 2005, p. 23). But essentialism – so long in the bad books amongst both biologists and philosophers – has been enjoying a strong resurgence. It all began with Kripke's classic *Naming and Necessity* (Oxford: Blackwell, 1972), although perhaps the laurel ought to go to Ruth Barcan-Marcus – see her 'Essentialism in Modal Logic' and 'Essential Attribution' in *Modalities: Philosophical Essays* (Oxford: Oxford University Press, 1993). See also Alvin Plantinga, *The Nature of Necessity* (Oxford: Clarendon Press, 1974); Hilary Putnam, *Mind, Language and World* (Cambridge: Cambridge University, 1975); Kit Fine, 'Postscript', in *Worlds, Times and Selves* in Fine and Prior (eds), (London: Duckworth, 1977); David Wiggins, *Sameness and Substance* (Oxford: Blackwell, 1980); and David Charles, *Aristotle on Meaning and Essence* (Oxford: Clarendon Press, 2000). For a general overview of contemporary formulations of essentialist theses see Graeme Forbes, 'Essentialism', in *A Companion to the Philosophy of Language*, Hale and Wright (eds). (Oxford: Blackwell, 1999, pp. 515–533).
5. The literature on Aristotle's metaphysics is very extensive and extraordi-narily sophisticated, and there is, unsurprisingly, room for rational debate regarding the details of his position. What I provide here, however, is relatively uncontroversial among Aristotle scholars. I follow the account given in Charles, op. cit. note 2.
6. It is not for nothing that Lawson-Tancred deemed Aristotle's 'the received metaphysics of the Western world'. In Aristotle, *Metaphysics*. Translation by Lawson-Tancred (London: Penguin, 2004, p. xxiii).

7. This argument is found in Ernst Mayr, 'Darwin and the Evolutionary Theory in Biology', *Evolution and Anthropology: A Centennial Appraisal*, Meggers (ed.), (Washington DC: Anthropology Society of Washington, 1959), and *The Growth of Biological Thought* (Cambridge MA: Belnap Press, 1982). It is also expounded in David Hull, 'The Effect of Essentialism on Taxonomy: Two Thousand Years of Stasis. Part 1', *Br. J. Philos. Sci.*, (1965), XVI: 1–18. See also M. T. Ghiselin (1981) 'Categories, Life and Thinking', *Behav. Brain Sci.*, (1981): 4, 269–283, 303–310.

8. This argument is also found in Mayr, op. cit. note 7.

9. That this latter point is required for the argument to have any force is not always spelled out explicitly, but John Dupré is clear on this. He doubts that descent is 'even a candidate for an essential property' because this property is 'purely relational'. *The Disorder of Things: Metaphysical Foundations of the Disunity of Science* (Cambridge, MA: Harvard University Press, 1993, p. 56).

10. For an expression of this argument see Samir Okasha, 'Darwinian Metaphysics: Species and the Question of Essentialism', *Synthese*, (2002): 131, 191–213.

11. See Eliot Sober, 'Evolution, Population Thinking, and Essentialism', *Phil. of Sci.*, (1980): 47, 350–383.

12. Op. cit., note 11, 370.

13. See M. Ereshefsky, 'Eliminative Pluralism', in *The Philosophy of Biology*, Ruse and Hul (eds), (Oxford University Press, 1998).

14. See R. Bernier, 'The Species as an Individual: Facing Essentialism', *Systematic Zoology*, (1984): 33 (4), 467.

15. 'Logical Difference and Biological Difference: The Unity of Aristotle's Thought', in *Philosophical Issues in Aristotle's Biology*, Lennox and Gotthelf (eds), (Cambridge: Cambridge University Press, 1987).

16. Balme, *Aristotle's de Partibus Animalium and De Generatione Animalium I* (Oxford: Clarendon Press, 1972).

17. Lennox, 'Material and Formal Natures in Aristotle's *de Partibus Animalium*' and 'Kinds, Forms of Kinds, and the More and the Less in Aristotle's Biology', in *Aristotle's Philosophy of Biology* (Cambridge: Cambridge University Press, 2001).

18. Walsh, 'Evolutionary Essentialism', *Brit. J. Phil. Sci.*, 57 (2006): 431.

19. Op. cit., note 18, 426.

20. Of course these arguments are redundant for those who already accept that species are natural groups, the foregoing reflections on the problem of change being sufficient to force essentialism.

21. That is, ancestral species A does not continue to exist in virtue of metamorphing into species B or C. Does this conflate sortal persistence conditions with diachronic identity conditions? Some metaphysicians want to distinguish the question 'Under what conditions can x remain the kind of thing x currently is?' from 'Under what conditions can x remain x?' Those who wish to preserve this distinction are motivated by the concern to allow for the possibility of metamorphosis of the sort associated with classical mythology, i.e., where Lucius, say, begins life

as a human being, is transformed into an ass and is ultimately returned to human form, all the while remaining Lucius. I think such scruples can be set aside here. For one, many will wonder whether the myths of metamorphosis are in fact fully intelligible (could Lucius really be an ass and remain Lucius?). For those whose intuitions prevent them from embracing metamorphosis as a genuine possibility sortal persistence conditions just are diachronic identity conditions because the identity of x is determined by x's sortal. But these considerations can be set aside in the current context because no evolutionary biologist believes that speciation events are cases of metamorphosis.

22. The crucial point about adaptations is that they are features or characters that *at some point* in their phylogenetic history were derived. That is, for a trait to be an adaptation there must have been at one stage of its history a transition from the ancestral to the derived state. This does not mean that this trait ceases to be an adaptation if it is subsequently passed on without modification to another species after further cleavage in the lineage. Adaptations can be, and often are, ancestral traits with respect to a particular set of species, say species C, D and E, where C is a daughter species of ancestral species A, and D and E are daughter species of C.

23. That phylogenetic trees are genuinely illuminating is assumed whenever they are employed in biochemistry, immunology, ecology, genetics, ethology, biogeography and stratigraphy. This assumption also underwrites a major methodological procedure in biology. Comparative analyses are only illuminating if the classification of the items being compared and their relationships are assumed to be accurate reflections of mind-independent biological reality. Thus phylogenetic trees taken to represent mind-independent biological reality are necessary for comparative anatomy, comparative physiology and comparative psychology. It is worth noting in this regard that realism about species is advocated by Darwin himself in the famous thirteenth chapter of *On the Origin of Species*. 'All the foregoing rules and aids and difficulties in classification are explained, if I do not greatly deceive myself, on the view that the natural system is founded on descent with modification; that the characters which naturalists consider as showing true affinity between any two or more species, are those which have been inherited from a common parent, and in so far, *all true classification is genealogical*; that community of descent is the hidden bond which naturalists have been unconsciously seeking, and not some unknown plan of creation, or the enunciation of general propositions, and the putting together and separating objects more or less alike'. *On the Origin of Species* in *From So Simple a Beginning. The Four Great Books of Charles Darwin*. E. O. Wilson (ed.), (New York: W.W. Norton, 2006, p. 717). If there is any question about how one is to read these lines, Darwin underlines his realism with the claim that 'This classification is evidently not arbitrary like the grouping of the stars in constellations' (op. cit., p. 711).

24. See Stearns and Hoekstra for further discussion (op. cit. note 1, 137).

25. Op. cit., note 9, 55.

26. *Developmental Plasticity and Evolution* (USA: Oxford University Press, 2003, p. 528).
27. Op. cit., note 27, 24.
28. *The Shape of Life: Genes, Development and the Evolution of Animal Form* (Chicago: University of Chicago Press, 1996, p. 31).
29. Op. cit., note 29, 360.
30. It is worth noting that geneticists are becoming increasingly comfortable with developmental programmes and using the notion to make discriminations between species. For an extended discussion of the biological details see chapter 6 of Stearns and Hoekstra's *Evolution: An Introduction*, 2nd Edition (Oxford: Oxford University Press, 2005). That developmental programmes might be the key to distinguishing biological species was raised over 30 years ago by King and Wilson in their 'Evolution at Two Levels in Humans and Chimpanzees' *Science*, 188, 1975: 107–166. Then the suggestion was used to account for the paradoxical fact that genetically human and chimpanzees are very similar while phenotypically very different. This suggestion has now received empirical support from various studies. The work of Khaitovich and Pääbo on primate gene expression offers a particularly clear example of how species specific variation in gene expression is now taken to be the distinguishing feature of a species. In particular see Philip Khaitovich, Wolfgang Enard, Michael Lachmann, Svante Pääbo, 'Evolution of Primate Gene Expression', *Nature Reviews Genetics*, 7, 2006: 693–702, Svante Pääbo, et al., 'Extension of Cortical Synaptic Development Distinguishes Humans from Chimpanzees and Macaques', *Genome Research*, published online February 2, 2012.
31. Some metaphysicians are willing to allow two objects to occupy the same space simultaneously. The standard example being a lump of clay and a vase composed of the clay. When the vase breaks the vase no longer exists but the clay remains, which means the vase was not the clay, and the clay was not the vase. One way to understand this is to maintain that the clay and the vase are two distinct objects which overlapped at one stage of their respective careers. But no one to my knowledge believes that this model can be extended to embrace the overlapping of two or more distinct organisms.

6 Evolutionary Biology, Modality and Explanation

1. This quote from Eddington is taken up enthusiastically by Fisher in the preface to his master piece *The Genetical Theory of Natural Selection* (1930, p. ix). It is echoed by Dawkins in *The Extended Phenotype* where he says '...to understand the actual we must contemplate the possible' (1999, p. 2).
2. Newton-Smith, particularly explicit about the state of play, writes: 'The current situation is an embarrassment for the philosophy of science. Indeed, one might go so far as to say that it is the sort of scandal to

philosophy of science that Kant thought scepticism was to epistemology. While we have insightful studies of explanation, we are a very long way from having this single unifying theory of explanation.…This task is made all the more pressing as most philosophers of science hold that *a* main task if not *the* main task, of science is to provide explanations' (2001, p. 132).

3. For an extended discussion of the standard approach to explanation in biology see Ernst Mayr's chapters 2 and 3 of his (1998).

4. For excellent discussions of both points see Ernst Mayr's *Systematics and the Origins of Species from the Viewpoint of a Zoologist* (Cambridge MA: Harvard University Press, 1942) and Paul Harvey and Mark Pagel's *The Comparative Method in Evolutionary Biology* (Oxford: Oxford University Press, 2000).

5. *Darwin's Dangerous Idea* (London: Penguin, 1995, p. 103).

6. The need for contrastive causes and alternative effects for determinate causal/explanatory relations seems to be accepted even by binary causal theorists given that they must posit a field of contrasts in order to identify causes.

7. Richard Dawkins, *The Blind Watchmaker* (London: Longmans, 1986, p. 73). Dawkins is not alone in this regard. Theodosius Dobzhansky writes: 'The variety of these possible ways of living – ecological niches is…great'. *The Genetics of the Evolutionary Process* (New York: Columbia University Press, 1970, p. 27).

8. McGhee, G. R., *Theoretical Morphology: The Concept and its Applications* (New York: Columbia University Press, 1999, p. 2).

9. Hickman, C. S., 'Theoretical Design Space: A New Program for the Analysis of Structural Diversity' in Seilacher and Chinzei (eds), *Progress in Constructional Morphology. Neues Jahrbuch fur Geologie und Palaontology, Abhandlungen*, 1990, p. 170.

10. (New York: Elsevier, 1988, p. xx). Simon Conway Morris sides with Lima-de-Faria on this point inasmuch as he takes certain evolutionary events to be virtually inevitable. See his *The Crucible of Creation: The Burgess Shale and the Rise of Animals.* (Oxford: Oxford University Press, 1998), and *Life's Solutions: Inevitable Humans in a Lonely Universe* (Cambridge: Cambridge University Press, 2004) for a perspective very much at odds with that expressed by Gould in *Wonderful Life: The Burgess Shale and the Nature of History* (London: Vintage, 2000).

11. The following set of contradictory theses about biological possibility and the explanatory power of natural selection is taken from Lima-de-Faria's *Evolution without Selection* (pp. 311–329). They illustrate the lack of professional consensus on biological modalities. The first in each pair is generally upheld by neo-Darwinian orthodoxy, the second by structuralists: 1. Biological evolution results from randomness and selection/ Biological evolution is totally conditioned by order found at the sub-atomic, elemental and mineral levels. 2. There are no physico-chemical constraints in gene and chromosome evolution/The genetic apparatus

was severely canalised when the gene and chromosome were formed. 3. Every type of cell, organism and trait is considered possible/Only certain types of cell organelles, of cells and of organisms were allowed to emerge. 4. Evolution has involved increased opportunities for variation/ Evolution created a restriction at every new step. 5. The number of forms is unlimited due to randomness and selection/The number of forms is limited and small, due to constraints imposed by the original construction of matter. The human smile is a function considered to be a legacy of natural selection conferring 'advantage'/the human smile defies interpretation by selection because it is automatic, like pain. As such it agrees with an interpretation based on the autoevolution of function.

12. This account of the intellectual division of labour is developed by E. J. Lowe in *The Possibility of Metaphysics* (Oxford: Oxford University Press, 2001).

13. The significance of this point should not be underplayed. As argued in Chapter 2, the best warrant one can provide for a metaphysical theory is precisely its ability to resolve puzzles of the sort identified earlier while preserving as much of the science as possible.

14. McGhee, G. R. *Theoretical Morphology: The Concept and Its Applications* (New York: Columbia University Press, 1999, p. 2).

15. There does not appear to be any purely geometrical constraint that would rule out the existence of cat-sized insects, for example. But since insects lack lungs, there are physical constraints on how much oxygen they can absorb, limits which constrain how large their bodies can become. So a morphospace that allowed for cat-sized insects would conflate the nonactual because it is physically impossible with the nonactual but biologically possible.

16. See Chapter 1 for the roots of modal eliminativism.

17. See David Lewis, *On the Plurality of Worlds* (Oxford: Blackwell, 1986).

18. See Alvin Plantinga, *The Nature of Necessity* (Oxford: Clarendon Press, 1974).

19. See Gideon Rosen, 'Modal Fictionalism', *Mind*, 99, 1990: 327–354.

20. Stephen Yablo, 'Is Conceivability a Guide to Possibility?' *Philosophical and Phenomenological Research*, 53, 1993: 1–42.

21. Williamson, Timothy, *The Philosophy of Philosophy* (Oxford: Blackwell, 2007, p. 163).

22. See chapter 3 of Robert Nozick's *Invariances* (Cambridge, MA: Harvard University Press, 2003). See also my 'The "Evolutionary Argument" and the Metaphilosophy of Common Sense', *Biology and Philosophy*, Vol. 22, no. 3, 2007: 369–382 for a detailed discussion of this line of thought. That our knowledge of what is possible depends on our knowledge of what is actual, or has been actualised, is itself a Scholastic principle of no small import. When discussing whether a real distinction obtains between two items in the natural order, Suarez maintains that there is a real distinction if the two items are separable. But can we recognise a real distinction between two items if we have no experience of those items

actually being separated? He writes: ' ... this precisely is the object of our inquiry – *how* can we know that they are separable, if they are not separated by natural means, and have not thus far been separated by God in any of the ways indicated?' (*Metaphysical Disputations*, 2007, VII, p. 49). The short answer is that we cannot.

23. Book IX of Aristotle's *Metaphysics* is the key source.

24. In Aristotelian terminology, potentiality follows on actuality (*Metaphysics*, Book IX, chapter 8), and actuality is determined by essence.

25. I am thinking in particular of standard experimental technique combined with inferential statistics. There are certain standard procedures the scientist engages in if she wishes to determine if there is a causal connection, rather than a mere association or correlation, between two items. Experiments bring together different objects under controlled conditions to see if any changes are brought about in the conditions of the items being investigated. If no changes are observed, the scientist must rule out the possibility that 'finks', 'antidotes' or 'masks' might be interfering in any possible causal relations that would otherwise obtain. If changes are observed, and these changes amount to correlations that are strong and linear, or associations that are 'statistically significant', then, if the experiment has been well designed (the sample size is large enough, the sample units have been chosen randomly and put randomly into control groups, etc.) the scientist is warranted in believing that a causal connection is the best explanation for the experimental results. It is probably unnecessary at this stage to say that this is how one investigates the natural order if one believes the world contains forms of non-logical necessity. Again the deep compatibility between Scholastic metaphysical principles and the sciences is in evidence. We have already had occasion to comment on the connection between operations and essences in Chapter 4 when dealing with the challenge of individuating living things. In that context we were concerned with the challenge of identifying individuals in order to count them. In our current context operations are a focus of attention because they reveal something of the nature of the individual.

26. For an introduction to some of the issues relating to the metaphysics of powers see Michael Esfeld's 'Humean Metaphysics vs a Metaphysics of Powers' in *Time, Change and Reduction: Philosophical Aspects of Statistical Mechanics*, in Ernst and Hütteman (eds) (Cambridge: Cambridge University Press, 2010, pp. 119–135). For more detailed discussions see *The Metaphysics of Powers*, Marmodoro (ed.) (London: Routledge, 2010), and *Properties, Powers and Structures*, Bird, Ellis and Sankey (eds), (London: Routledge, 2011).

27. See also Rejane Bernier, 'The Species as Individual: Facing Essentialism', *Systematic Zoology*, 33 (4), 1984: 460–469, Dennis Walsh, 'Evolutionary Essentialism', *Brit. J. Phil.Sci.* 57, 2006: 425–448 and Michael Devitt, 'Resurrecting Biological Essentialism', *Philosophy of Science*, 75, 2008: 344–382.

28. There are two ways in which x may fall within the scope of a developmental programme. The trajectory of a developmental programme is often affected by factors external to the organism. Certain environmental factors can influence the activity of control genes. Now if the actualisation of such external factors would lead a developmental programme down a particular path to x, then x falls within the scope of the developmental programme, and is biologically possible. But the trajectory of a developmental programme can be altered by a mutation on a control gene or by experimental perturbation. (The duplication of control genes is perhaps the most common mutation that opens up new developmental trajectories; manipulation of development by using colchicine to affect the activity of control genes is one example of experimental perturbation.) As such mutations arise in the natural course of things, and particularly because we can now intervene in developmental processes, all that is required for the change in developmental programmes currently exists. This gives us a second sense in which x can fall within the scope of a developmental programme. If a single mutation on or perturbation of a control gene would lead a developmental programme down a particular path to x, then x falls within the scope of that programme, and is biologically possible.

7 Evolutionary Biology and Ethics

1. See Rolston's (1988) for discussion.
2. This is how Kitcher (1993) understood Ruse and Wilson (1986).
3. Again, Kitcher (1993) takes Ruse and Wilson to task on precisely this claim.
4. Here I take issue with Sharon Street (2006) who has argued that some versions of moral realism are incompatible with the Darwinian view that our evaluative attitudes are themselves the products of natural selection.
5. Moore's *Principia Ethica* is the source of this argument.
6. Rose (1978) nicely captures the anxieties associated with genetic determinism. See Dawkins (1999, chapter 2) and Wilson (2004, chapter 4) for attempts by non-philosophers to cope with the threat of genetic determinism.
7. Wright (1994, p. 326) ends a chapter on 'Darwinian cynicism' with the following summary indicative of this line of argument: 'Thus the difficult question of whether the human animal can be a moral animal – the question that modern cynicism tends to meet with despair – may seem increasingly quaint. The question may be whether, after the new Darwinism takes root, the word *moral* can be anything but a joke'.
8. The firm denial of progress in evolution, apart from its value as a statement of empirical fact, is usually employed to counter unwelcome associations with Spencer. Spencer saw cultural progress as simply an

extension of organic progress, and argued that the forces of organic progress, i.e., natural selection, ought to be encouraged in the cultural sphere. He wrote: '...we propose in the first place to show that this law of organic progress is the law of all progress. Whether it be in the development of the Earth, in the development of Life upon its surface, in the development of Society, of Government, of Manufactures, of Commerce, of Language, Literature, Science, Art, this same evolution of the simple into the complex, through successive differentiations, holds throughout' (in Ruse, 2009, p. 20). The insistence that natural selection ought to be allowed to operate unchecked in the social order led to Spencer's unsavoury statements which still haunt sociobiology to this day. The language used in his condemnation of 'spurious philanthropists' is harrowing: 'Blind to the fact that under the natural order of things, society is constantly excreting its unhealthy, imbecile, slow, vacillating, faithless members, these unthinking, though well-meaning, men advocate an interference which not only stops the purifying process but even increases the vitiation' (in Ruse, 2009, p. 69).

9. See Sharron Street (2006) for an extended presentation of this line of thought.

10. Weber writes: 'The fate of our times is characterised by rationalisation and intellectualisation and, above all, by "the disenchantment of the world". Precisely the ultimate and most sublime values have retreated from public life into either the transcendental realms of mystic life or into the brotherliness of direct and personal human relations' (in 'Science as a Vocation', in Gerth and Mills (1946, pp. 129–156)).

11. The following account of Weber relies heavily on Michael Lessnoff (1999, chapter 2, 'Max Weber and the Politics of the Twentieth Century').

12. I take the core commitments of classical Western society to be (a) a reliance upon the natural sciences as the primary source of knowledge of the workings of the natural order; (b) a reliance upon some form of representative democracy and the rule of law as the chief means by which we arrive at decisions concerning our collective actions; and (c) a reliance upon market economics as the primary means of determining what non-public goods and services are produced, in what quantities and at what price. In moments of crisis we are likely to reconsider our commitment to all three of these core stances, as recent history has shown only too well.

13. This is the so-called agent, as opposed to 'act', centred approach to ethics.

14. Epicurus writes: 'Empty is that philosopher's argument by which no human suffering is therapeutically treated. For just as there is no use in a medical art that does not cast out the sicknesses of the body, so too there is no use in philosophy, unless it casts out the suffering of the soul' (in Nussbaum, 1994, p. 13). See Nussbaum (1994) for an extended study of the medical conception of ethics as found in the Ancient Greeks. See also Lloyd (2003) for a very useful study placing the ethical thinking of Plato and Aristotle within the broader context of Ancient Greek thought

on health and disease. Llyod points out that (a) Aristotle develops Plato's comparison between the state and the individual, with the health of the individual organism serving as the model of well-being for both individuals and the state; that (b) Aristotle makes use of the notions of health and disease to support moral realism, and to insist that, just as there are medical experts, so too there can be ethical experts; and (c) that Aristotle uses the notions of health and disease to evaluate political constitutions (2003, p. 176). It is worth noting as well that this conception of ethics seems to have occurred to Darwin himself. He writes: 'The term, general good, may be defined as the term by which the greatest number of individuals can be reared in full vigour and health, with all their faculties perfect, under the conditions to which they are exposed'.

15. According to the latest HDI report the top ten countries in the world in terms of human development are: Norway, Australia, The United States, The Netherlands, Germany, New Zealand, Ireland, Sweden, Switzerland and Japan. The bottom ten countries are: Burundi, Guinea, Central African Republic, Eritrea, Mali, Burkina Faso, Chad, Mozambique, The Congo, and Niger. On the model of ethics proposed here the main task is to understand the causes behind the different levels of developmental attainment to then identify policy priorities for developing countries.

16. The best response to those who would downplay the role of health in ethical reflection is that they are simply not playing the ethics game as understood by those who adopt the medical conception of the discipline.

17. For instance, many Western governments (the United Kingdom, Germany and the United States) as well as Japan managed to improve the standard of living of their citizens by focusing on policies that prioritised the creation of a national transportation system, a national banking system (to stabilise the currency and lend to businesses), a system of mass education and the elimination of tariffs within the national borders.

18. The policies mentioned in the previous note have been exported to other countries, not always with equal success. Identifying the modifications required to achieve development in standards of living is part of the challenge of developmental economics.

19. Works like Amartya Sen's *Development as Freedom*, and Partha Dasgupta's *An Inquiry into Well-Being and Destitution* deserve a much wider audience amongst philosophers and ethicists.

20. For discussion of this line of thought see my (2007), chapters 2 and 8.

Bibliography

Aquinas, T. (1986). *The Divisions and Methods of the Sciences.* Toronto, The Potinfical Institute of Medieval Studies.

—— (1995). *Commentary on Aristotle's Metaphysics.* Notre Dame, Dumb Ox Books.

—— (1997). *Basic Writings of St Thomas Aquinas.* Indianapolis, Hackett Publishing Company. 2.

Aristotle (1941). *The Basic Works of Aristotle.* New York, Random House.

—— (2004). *Metaphysics.* London, Penguin.

Autrecourt, N. (1971). *The Universal Treatise.* Translation by Kennedy, Arnold and Millward. Milwaukee, Marquette University Press.

Ayala, F. (2010). 'What the Biological Sciences Can and Cannot Do for Ethics' in Ayala and Arp (eds). *Contemporary Debates in Philosophy of Biology.* Oxford, Blackwell, 316–336.

Balme, D. (1972). *Aristotle's de Partibus Animalium and De Generatione Animalium I.* Oxford, Clarendon Press.

Barcan-Marcus, R. (1993). *Modalities: Philosophical Essays.* Oxford, Oxford University Press.

Bermudez, J. L. (2000). 'The Originality of Cartesian Scepticism: Did It Have Ancient or Medieval Antecendents?' *History of Philosophy Quarterly* **17**: 333–360.

Bernier, R. (1984). 'The Species as Individual: Facing Essentialism.' *Systematic Zoology* **33** (4): 460–469.

Bird, Ellis. and Sankey, Ed. (2011). *Properties, Powers and Stuructures.* London, Routledge.

Bosley, R. and Tweedale M., Eds (2006). *Basic Issues in Medieval Philosophy.* Peterborough, Broadview Press.

Boulter, S. (2007). *The Rediscovery of Common Sense Philosophy.* Houndmills, Palgrave Macmillan.

Boulter, S. and Boulter, S. (2011). 'The Medieval Origins of Conceivability Arguments.' *Metaphilosophy* **42** (5): 617-641.

Boulter, S. (2012). 'Can Evolutionary Biology do without Aristotelian Essentialism?' *Human Nature: Royal Institute of Philosophy Supplement* **70**: 83–103.

Boulter, S. (2012). 'Contrastive Explanations in Evolutionary Biology. *Ratio* XXV (4): 425–441.

Boulter, S (2012). 'The Aporetic Method and the Defense of Immodest Metaphysics.' in Edward Feser (ed.). Aristotle on Method and Metaphyics. Houndsmill, Palgrave Macmillan.

Boulter, S. (2013). 'Aquinas on Biological Individuals: An Essay in Analytical Thomism.' Philosophia.

Buridan, J. (2001). Summa de Dialectica. Translation by Klima, G. New Haven, Yale University Press.

Carnap, R. (1950). 'Empiricism, Semantics and Ontology.' Revue Internationale de Philosophie IV: 20–40.

Chalmers, D. (1996). *The Conscious Mind*. New York, Oxford University Press.

Charles, D. (2000). *Aristotle on Meaning and Essence*. Oxford, Clarendon.

Clark, E. (2010) 'The Problem of Biological Individuality', Biological Theory 5 (4): 312–325.

Colodny, Ed. (1962). *Frontiers of Science and Philosophy*. Pittsburgh, Pittsburgh University Press.

Darwin, C. (2006). *From So Simple a Beginning: The Four Great Books of Charles Darwin*. New York, W.W. Norton.

Dasgupta, P. (1993). *An Inquiry into Well-Being and Destitution*. Oxford, Clarendon Press.

David Chalmers, D. M. and Ryan Wasserman, Ed. (2009). *Metametaphysics: New Essays on the Foundations of Ontology*. Oxford, Oxford University Press.

Dawkins, R. (1986). *The Blind Watchmaker*. London, Longmans.

—— (1999). *The Extended Phenotype*. Oxford, Oxford University Press.

Dennett, D. (1995). *Darwin's Dangerous Idea*. London, Penguin.

Descartes, R. (1990). *The Philosophical Writings of Descartes*. Cambridge, Cambridge University Press. **1**.

—— (2008). *The Philosophical Writings of Descartes*. Cambridge, Cambridge University Press. **2**.

Devitt, M. (2008). 'Resurrecting Biological Essentialism.' *Philosophy of Science* **75**: 344–382.

Dobzhansky, T. (1949). *Genetics and the Origin of Species*. New York, Columbia University Press.

—— (1970). *The Genetics of the Evolutionary Process*. New York, Columbia University Press.

Duhem, P. (1977). *The Aim and Sturucture of Physical Theory*. New York, Athenium.

Dupre, J. (1993). *The Disorder of Things*. Cambridge, MA, Harvard University Press.

Eddington, A. S. (1928). *The Nature of the Physical World*. New York, Macmillan.

Ernst and Hutteman, Eds (2010). *Time, Change, and Reduction: Philosophical Aspects of Statistical Mechanics*. Cambridge, Cambridge University Press.

Esfeld, M. (2010). 'Humean Metaphysics vs. a Metaphysics of Powers' in Ernst and Hutteman , 119-135.

Feser, E., Ed. (2013). *Aristotle on Method and Metaphysics*. Houndmills, Palgrave Macmillan.

Fisher, R. A. (1930). *The Genetical Theory of Natural Selection*. Oxford, Clarendon Press.

Fine, K. (1977). 'Postscript' in Fine and Prior.

Fine, K and Prior, A., Eds (1977). *World, Time and Selves*. London, Duckworth.

Frede, M. (1988). 'A Medieval Source of Modern Scepticism.' *Gedankenzeichen: Festschrift fur Klaus Oehler zum* **60**: 67–70.

Galilei, G. (2001). *Dialogue Concerning on the Two Chief World Systems: Ptolomeic and Copernican.* New York, Modern Library.

Gellner, E. (2006). *Nations and Nationalism.* 2nd edition. Oxford, Blackwell.

Gendler, A. and Hawthorne, T., Eds (2002). *Conceivability and Possibility.* Oxford, Oxford University Press.

Gerth and Mills, Eds (1946). *From Max Weber: Essays in Sociology.* New York, Oxford University Press.

Ghent, H. O. (2011). *Summa of Ordinary Questions: Articles Six to Ten on Theology.* Milwaukee, Marquette University Press.

Ghiselin, M. T. (1981). 'Categories, Life and Thinking.' *Behavioral and Brain Science* **4**: 269–283, 303–310.

Gilson, E., Ed. (1937). *The Unity of Philosophical Experience.* New York, Charles Scribner's Sons.

Gould, S. J. (2000). *Wonderful Life: The Burgess Shale and the Nature of History.* London, Vintage.

Griffiths, S. a. (1999). *Sex and Death: An Introduction to Philosophy of Biology.* Chicago, University of Chicago Press.

Gutting, G. (2009). *What Philosophers Know: Case Studies in Recent Analytical Philosophy.* Cambridge, Cambridge University Press.

H., R. I. (1988). *Environmental Ethics: Duties to and Values in the Natural World.* Philadelphia, Temple University Press.

Hand, J. D. (2008). *Statistics: A Very Short Introduction.* Oxford, Oxford University Press.

Harrmann, N. (1965). *Grundzuge einer Metaphysik der Erkenntnis.* Berlin, W. de Gruyter.

Harvey and Pagel (2000). *The Comparative Method in Evolutionary Biology.* Oxford, Oxford University Press.

Hauser, M. (2006a). 'The Liver and the Moral Organ.' *Social Cognitive and Affective Neuroscience Advance Access*: 1 (3): 314–220.

—— (2006b). *Moral Minds.* London, Abacus.

Hickman, C. S. (1990). 'Theoretical Design Space: A New Program for the Analysis of Structural Diversity.' *Progress in Constructional Morphology. Neues Jahrbuch for Geologie und Palaontology, Abhandlungen 190*: 169–182.

Hobbes, T. (1985). *Leviathan.* London, Penguin.

Hill, B. and Lagerlund, H., Eds (2012). *The Philosophy of Francisco Suarez.* Oxford, Oxford University Press.

Hull, D. (1965). 'The Effect of Essentialism on Taxonomy: Two Thousand Years of Stasis. Part I.' *British Journal of the Philosophy of Science* **26**: 1–18.

Hull, D and Ruse, M., Eds (1998). *The Philosophy of Biology.* Oxford, Oxford University Press.

Hume, D. (1989). *A Treatise of Human Nature.* Oxford, Clarendon Press.

Haeckel, E. (1979). *Das System der Medusen.* Jena. Gustav Fischer.

Hull, D. (2001). *Science and Selection: Essays on biological evolution and the philososphy of science.* Cambridge, Cambridge University Press.

Huxley, J. (1912). *The Individual in the Animal Kingdom.* Cambridge, Cambridge University Press.

Huxley, T. H. (1852). 'Upon Animal Individuality' in Proceedings of the Royal Institute, April 30, 184–189.

Inwood, B. and Gerson, L. P. (1997). *Hellenistic Philosophy: Introductory Readings.* Indianapolis, Hackett Publishing Company.

Kant, I. (1953). *Groundwork of the Metaphysics of Morals.* Translation by H.J. Paton as The Moral Law. London, Hutchison.

Irwin, T. (1974). *Aristotle's First Principles.* Oxford, Clarendon Press.

Jackson, D. B.-M. A. F. (2007). *Philosophy of Mind and Cognition: An Introduction.* Oxford, Blackwell.

Kant, I. (1965). *Critique of Pure Reason.* New York, St Martin's Press.

Kirk, R. a, S. (1991). *The Presocratic Philosophers.* Cambridge, Cambridge University Press.

Kitcher, P. (2009). 'Four Ways of Biologizing Ethics' in Ruse, M., 379–388.

Klima, G. (2007). *Medieval Philosophy.* Oxford, Blackwell.

Kripke, S. (1972). *Naming and Necessity.* Oxford, Blackwell.

Leibniz, G. W. (1983). *Philosophical Writings.* London, Everyman's Library.

Lennox (2001). *Aristotle's Philosophy of Biology.* Cambridge, Cambridge University Press.

Lennox, G. A., Ed. (1987). *Philosophical Issues in Aristotle's Biology.* Cambridge, Cambridge University Press.

Lessnoff, M. (1999). *Political Philosophers of the Twentieth Century.* Oxford, Blackwell.

Lewis, D. (1986). *On the Plurality of Worlds.* Oxford, Blackwell.

—— (1987). *On the Plurality of Worlds.* Oxford, Basil Blackwell.

Lima-de-Faria (1988). *Evolution without Selection.* New York, Elsevier.

Lowe, E. J. (2001). *The Possibility of Metaphysics.* Oxford, Oxford University Press.

Lloyd, G. E. R. (2003). In the Grip of Disease. Oxford, Oxford University Press.

—— (2002). *A Survey of Metaphysics.* Oxford, Oxford University Press.

—— (2007) *The Four-Category Ontology: A Metaphysical Foundation for Natural Science.* Oxford, Oxford University Press.

Markowski, M. (1984). 'L'influence de Jean Buridan sur les universities d'Europe centrale' in Kaluza, Z. and Vignaux, P. (eds). *Preuves et raisons a l'universite de Paris.* Paris, Vrin, 149–163.

Marmodoro, Ed. (2010). *The Metaphysics of Powers.* London, Routledge.

Massey, H. a., Ed. (1991). *Thought Experiments in Science and Philosophy.* Lanham, Rowan and Littlefield.

Mayr, E. (1942). *Systematics and the Origins of Species from the Viewpoint of a Zoologist.* Cambridge, MA, Harvard University Press.

—— (1982). *The Growth of Biological Thought.* Cambridge, MA, Harvard Unviersity Press.

—— (1998). *This Is Biology: The Science of the Living World.* Cambridge, MA, Belnap Press.

McGhee, G. R. (1999). *Theoretical Morphology: The Concept and Its Applications.* New York, Columbia University Press.

Meggers, Ed. (1959). *Evolution and Anthropology: A Centennial Appraisal.* Washington DC, Anthropology Society of Washington.

Moore, G. E. (1963). *Philosophical Papers.* New York, Macmillan.

—— (1993). *Principia Ethica.* Cambridge, Cambridge University Press.

Morris, S. C. (1998). *The Crucible of Creation: The Burgess Shale and the Rise of Animals.* Oxford, Oxford University Press.

—— (2004). *Life's Solutions: Inevitable Humans in a Lonely Universe.* Cambridge, Cambridge University Press.

Nagel, T. (1974). 'What Is It Like to Be a Bat?' *Philosophical Review* **83**: 435–450.

Newton-Smith, W. H. (2001). 'Explanation' in Newton-Smith (ed.). *A Companion to the Philosophy of Science.* Oxford, Blackwell, 127–132.

Nietzsche, F. (1966). *Beyond Good and Evil.* New York, Vintage Books.

Nussbaum, M. (1994). *The Therapy of Desire.* Princeton, Princeton University Press.

Ockham, W. (1983). *Tractatus de praedestinatione et de prasescientia Dei et de futuris contingentibus.* Indianapolis, Hackett Publishing Company.

—— (1990). *Ockham: Philosophical Writings.* Indianapolis, Hackett Publishing Company.

—— (1991). *Quodlibetal Questions.* New Haven, Yale University Press.

Oderberg, D. (2007). *Real Essentialism.* London, Routledge.

Okasha, S. (2002). 'Darwinian Metaphysics: Species and the Question of Essentialism.' *Synthese* **131**: 191–213.

Pagel, P. H. A. M. (2000). *The Comparative Method in Evolutionary Biology.* Oxford, Oxford University Press.

Pasnau, R., Ed. (2002). *The Cambridge Translation of Medieval Philosophical Texts.* Cambridge, Cambridge University Press.

Pepper, J. and heron, M. (2008). 'Does Biology Need an Organism Concept?' *Biological Reviews* 83 (4): 621–627.

Philip Khaitovich, W. E., Michael Lachmann and Svante Paabo (2006). 'Evolution of Primate Gene Expression.' *Nature Reviews Genetics* **7**: 693–702.

Pinborg, J. (1976). *The Logic of John Buridan: Acts of the European Symposium on Medieval Logic and Semantics.* Copenhagen, Museum Tusculanum.

Plamenatz, J. (1992). *Man and Society: Political Theories from Machiavelli to Marx.* London, Longman. **1**.

Plantinga, A. (1974). *The Nature of Necessity.* Oxford, Clarendon Press.

Plato (1961). *Plato: Collected Dialogues.* Princeton, Princeton University Press.

Prior, F. A., Ed. (1977). *Worlds, Times and Selves.* London, Duckworth.

Putnam, H. (1975). *Mind, Language and World.* Cambridge, Cambridge University Press.

Quine, W. V. O. (1953). *From a Logical Point of View.* Cambridge, MA, Harvard University Publishing.

Raff, R. (1996). *The Shape of Life: Genes, Development and the Evolution of Animal Form.* Chicago, University of Chicago Press.

Rea, M., Ed. (2009). *Arguing about Metaphysics.* London, Routledge.

Rescher, N. (2009). *Aporetics: Rational Deliberation in the Face of Inconsistency.* Pittsburgh, Pittsburgh University Press.
—— (2006). *Philosophical Dialectics.* Albany, SUNY.
Rolston, H. (1988). *Environmental Ethics: Duties to and Values in the Natural World.* Philadelphia, Temple University Press.
Rose, S. (1978). 'Pre-Copernican Sociobiology?' *New Scientist* 80: 45-46.
Rosemann, P. (2004). *Peter Lombard.* Oxford, Oxford University Press.
Rosen, G. (1990). 'Modal Fictionalism.' *Mind* 99: 327–354.
Ruis-Mirazo, K. et al. (2000). 'Organisms and Their Place in Biology.' *Theory in Biosciences* 118 (3), 209–233.
Ruse, H. A., Ed. (1998). *The Philosophy of Biology.* Oxford, Oxford University Press.
Ruse, M. (2010). 'The Biological Sciences Can Act as a Ground for Ethics' in Ayala and Arp (eds). *Contemporary Debates in Philosophy of Biology.* Oxford, Blackwell, 297–315.
Ruse, M. (2009). *Philosophy After Darwin.* Princeton, Princeton University Press.
Ruse, M. and Wilson, E. O. (1986). 'Moral Philosophy as Applied Science.' *Philosophy* 61: 173–192.
Ryle, G. (2002). *Dilemmas.* Cambridge, Cambridge University Press.
Scotus, D. (1949). *De Primo Principio.* St Bonaventure, The Fransican Institute.
Sen, A. (1981). *Poverty and Famines.* Oxford, Clarendon Press.
Sen, A. (1999). *Development as Freedom.* Oxford, Oxford University Press.
Sober, E. (1980). 'Evolution, Population Thinking and Essentialism.' *Philosophy of Science* 47: 350–383.
Sorensen, R. (1992). *Thought Experiments.* Oxford, Oxford University Press.
Spinoza, B. (1998). *Principles of Cartesian Philosophy.* Indianapolis, Hackett Publishing Company.
Stachel, J. (2002). *Einstein from 'B' to 'Z'.* Boston, Birkhauser.
Stearns and Hoekstra (2005). *Evolution: An Introduction.* Oxford, Oxford University Press.
Sterelny and Griffiths (1999). *Sex and Death: An Introduction to Philosophy of Biology.* Chicago, University of Chicago Press.
Stevenson, L. (1987). *Seven Theories of Human Nature.* Oxford, Oxford University Press.
Street, S. (2006). 'A Darwinian Dilemma for Realist Theories of Value.' *Philosophical Studies* 127: 109–166.
Suarez, F. (1983). *On the Essense of Finite Being as such, on the Existence of That Essence and Their Distinction.* Milwaukee, Marquette University Press.
Suarez, F. (1964). *Metaphysical Disputations VI: On Formal and Universal Unity.* Translation by Ross. Milwaukee, Marquette University Press.
—— (2007). *Metaphyhsical Disputations: On the Various Kinds of Distinction.* Milwaukee, Marquette University Press.
Suppe, F., Ed. (1974). *The Structure of Scientific Theories.* Urbana, University of Illinois Press.

Svante Paabo, P. K. E. A. (2012). 'Extension of Cortical Synap Development Distinguishes Humans from Chimpanzees and Macaques.' *Genome Research* 22: 611–622.

Takho, T., Ed. (2012). *Contemporary Aristotelian Metaphysics.* Cambridge, Cambridge University Press.

Unger, P. (2009). 'I do not Exist' in M. Rea (ed.). *Arguing about Metaphysics.* London, Routledge, 94–105.

Vignaux, K. A., Ed. (1984). *Preuves et raisons a l'universite de Paris.* Paris, Vrin.

Walsh, D. (2006). 'Evolutionary Essentialism.' *British Journal of the Philosophy of Science* 57: 425–448.

Weber, M. (1946). 'Science as a Vocation' in Gerth and Mills, 129–156.

West-Eberhard, M. J. (1998). 'Adaptation: Current Usages' in Hull and Ruse, 8–14.

West-Eberhard, M. J. (2003). *Developmental Plasticity and Evolution.* New York, Oxford University Press.

Wiggins, D. (1980). *Sameness and Substance.* Oxford, Blackwell.

Williamson, T. (2007). *The Philosophy of Philosophy.* Oxford, Blackwell.

Wilson, E. O. (2004). *On Human Nature.* Cambridge, MA, Harvard University Press.

Wilson, J. (1999). *Biological Individuality: The Identity and Persistence of Living Things.* Cambridge, Cambridge University Press.

Wilson, K. a. (1975). 'Evolution at Two Levels in Humans and Chimpanzees.' *Science* 188: 107–166.

Wittgenstein, L. (1986). *Tractatus Logico-Philosophicus.* London, Routledge and Kegan Paul.

Wright, R. (1994). *The Moral Animal.* New York, Vintage Books.

Wright, C. and Hale, B., Eds (1999). *A Companion to the Philosophy of Language.* Oxford, Blackwell.

Wuellner, B. (1956). *Summary of Scholastic Principles.* Chicago, University of Loyola Press.

Yablo, S. (1993). 'Is Conceivability a Guide to Possibility?' *Philosophical and Phenomenological Research* 53: 1–42.

Yoltan, J. (1983). *Thinking Matter: Materialism in the Eighteenth-Century.* Minneapolis, University of Minnesota Press.

Zalta, E. (1983). *Abstract Objects.* Dordrecht, D. Reidel Company.

Zimmerman, V. I. A., Ed. (1998). *Metaphysics: The Big Questions.* Oxford, Blackwell.

Zupka, J. (1993). 'Buridan and Scepticism.' *Journal of the History of Ideas* 31: 191–221.

Index

Printed in the United States
by Baker & Taylor Publisher Services